Essential Aspects of Mass Spectrometry

Essential Aspects of Mass Spectrometry

Alberto Frigerio
Chief, Laboratory of Mass Spectrometry
Mario Negri Institute for Pharmacological Research
Milan, Italy

Edited by
Roy Carrington
Welcome Trust Research Fellow
Mario Negri Institute for Pharmacological Research

SPECTRUM PUBLICATIONS, INC.
Flushing, New York

Distributed by Halsted Press
A Division of John Wiley & Sons
New York   Toronto   London   Sydney

Spectrum Publications, Inc.
75-31 192 Street, Flushing, New York 11366

Distributed solely by the Halsted Press Division of John Wiley & Sons, Inc., New York.

Library of Congress Cataloging in Publication Data

Frigerio, Alberto.
Essential aspects of mass spectrometry.

Bibliography: p.
1. Mass spectrometry. I. Title.
QC454.M3F75 545′.33 74–12348
ISBN 0–470–28120–0

FOREWORD

It has been estimated recently that during the last 50 years mankind has produced as many scientists as in all its previous historical evolution. Obviously, depending on natural laws, this means that at least 80% of such a particular population are still alive and working.

Apart from any positive or negative evaluation of this fact such a situation has brought about an explosive increase in human knowledge and this gives rise to two main problems which directly involve the scientist. The first problem is the increasing difficulty that the scientist finds to be adequatedly acquainted with the manifold aspects and the continuous extension of science, often in his own field of research. The other problem, which makes the former situation worse, is the huge amount and spreading proliferations of very sophisticated techniques devised to deepen the search for scientific truth.

In addition, it has to be considered that some of these techniques have reached such an individuality and complexity that they may be regarded as new branches of science.

Certainly this is the case of mass spectrometry, which requires a deep technical preparation and a very flexible attitude to be usefully and properly applied to the various scientific demands.

For many years I have been involved in the manifold aspects of neurosciences and (I am afraid) I must confess to be very poorly acquainted with mass spectrometry and its applications. So, as a direct consequence of my ignorance of the field, when Doctor Frigerio gave me his book, I approached it with a little mistrust but, after having read the first few pages I immediately became interested, losing most of my inhibitions towards a field which I believed almost uncomprehensible to me.

This fact may be taken as evidence of Doctor Frigerio's capability to present and develop, in a very clear and simple manner, a most complex and difficult matter.

Mass spectrometry has become of prime importance for the solution of many scientific problems and, presumably, it will become essential for the forthcoming developments of science. With such a prospect in view, I think that this volume by Dr. Frigerio should be included in that precious category of works which constitutes the ground upon which one can either build up his specialization or enlarge his purview.

Luigi Valzelli, M.D.
Head Section of Neuropsychopharmacology
Istituto di Ricerche Farmacologiche
"Mario Negri"
Milan, Italy

PREFACE

In the last few years mass spectrometry has become an analytical method of prime importance in the identification and structural analysis of organic compounds. For some problems, such as the determination of components separated by gas chromatography, it is one of the few techniques able to give specific data from a sample at the nanogram level. One of its major limitations is that it requires a vaporized sample.

Mass spectrometry has become an indispensible tool in organic chemistry, pharmacology, biochemistry, medicine, toxicology, food chemistry, forensic science, petrology, geochemistry, pollution studies, and many other sectors of research.

This book has been written to give students in chemistry and other scientific disciplines a general outline of the use of this spectrometric method. It gives the reader the fundamental principles necessary for the understanding of the fragmentation mechanisms and the interpretation of a mass spectrum.

It should be treated therefore as an introductory book. It does not pretend to be original in all parts, but I have attempted to give a systematic explanation of the basic aspects of this technique, and I hope the book may be of value in this respect. I think that this volume may also arouse interest in those who, although no longer students, come in contact with problems involving mass spectrometry and may desire information on the essential aspects of this technique.

I would like to thank Professor Bruno Danieli for his helpful comments in the preparation of the manuscript, and I am grateful to Professor Luigi Valzelli for his kind foreword.

Milan, Italy
May 30, 1974

Alberto Frigerio

CONTENTS

CHAPTER I

A. INTRODUCTION.

The fundamental principle of mass spectrometry, the separation and recording of the mass of an ionized atom, was demonstrated many years ago.

In 1898, W. Wien deflected a beam of positive ions in electric and magnetic fields and, in 1912, J. J. Thompson was able to demonstrate the existence of two isotopes of neon, masses 20 and 22, using a magnetic deflection instrument.

The first precision instruments were constructed by J. Dempster in 1918 and by F. W. Aston in 1919, to measure the relative abundances of some isotopes. Aston's spectrograph was particularly useful for the accurate measurement of mass because the ions were focused onto a photographic plate (hence the term "spectrograph").

Until 1940 the mass spectrometer was only used for the analysis of gases and for the determination of the stable isotopes of chemical elements. It was later used to carry out quick and accurate analyses of complex mixtures of hydrocarbons from petroleum fractions; then, when it was demonstrated that a complex molecule could give rise to a well-defined and reproducible mass spectrum, interest in its application to the determination of organic structures was established.

At present, the mass spectrometer is an indispensible tool in organic chemistry, pharmacology, biological chemistry, toxicology, petrochemicals, geochemistry, pollution studies, and in many other types of research.

The coupling of the gas chromatograph with mass spectrometry has extended the applications of both techniques. These two methods are, in fact, highly complementary: The gas chromatograph is efficient in the separation of the constituents of a mixture but does not always give a full identification, whereas the mass spectrometer can identify a single compound but is less efficient in the study of complex mixtures. A very important factor is the comparable sensitivity of the two techniques.

One of the advantages of the mass spectrometer is that it can record spectra from small quantities of substance. In fact, the technique gives more information for a nanogram (10^{-9}g) of sample than any other method at our disposal.

In organic and biological experiments, very small quantities of very precious substances are often isolated, from natural sources, after accurate and patient purification by thin layer chromatography. The only possible method of analysis, in such cases, is mass spectrometry.

With a careful study of the spectrum and with the help of a standard, it is possible to arrive at the structure of the product.

B. INSTRUMENTATION.

The mass spectrometer is an instrument that produces ions from a molecule, separates the ions as a function of their mass-to-electric charge ratio ($^m/e$), and records and displays the relative abundance of these ions.

Although it is possible to study both positive and negative ions with a mass spectrometer, the major use of the instrument is in the study of positive ions. Throughout the following pages, the term "ions" will always refer to positive ions.

In principle, the function of the mass spectrometer is relatively simple. The molecules of the substance, in a gaseous phase, become ionized, the ions produced are accelerated in an electric field of high potential, they become deviated in a magnetic field and then arrive at a collector, generating a signal the intensity of which is proportional to the number of ions arriving. The whole apparatus operates under high vacuum.

The record of the signals constitutes the mass spectrum.

To illustrate the function of a mass spectrometer (figure 1), imagine a stone being projected from a catapult toward a delicate vase (I). On impact the vase is shattered (II). If the pieces are carefully collected (III), the vase can be reconstructed from the fragments (IV).

In this example the vase represents the molecule, the catapult the filament, and the stone the bombarding electron.

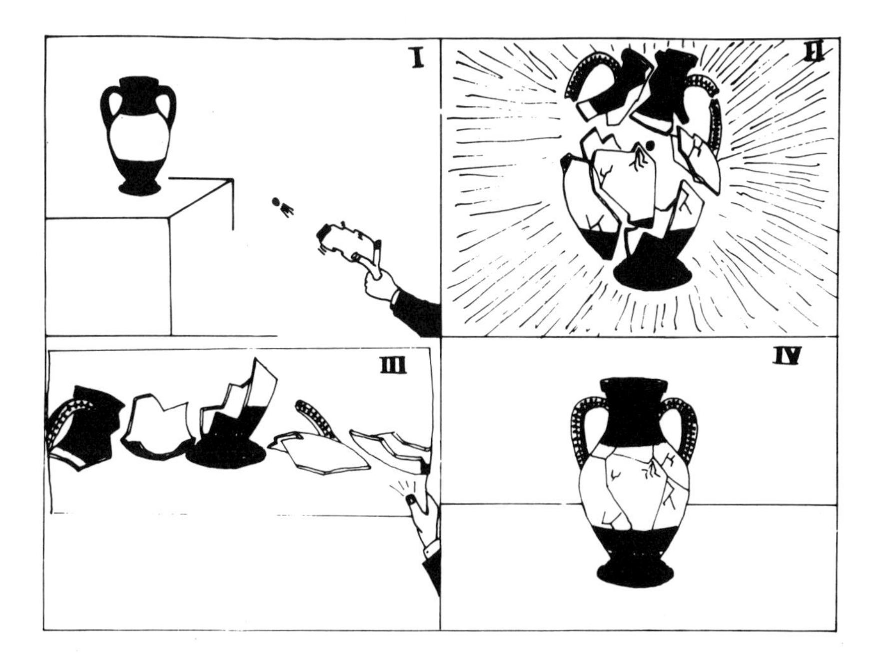

Figure 1 Illustration of the principle of mass spectrometry by electron bombardment.

next 6

The different types of recorder employed distinguish the "spectrograph" from the "spectrometer"; otherwise they are substantially the same. In the mass spectrograph the ions are focused onto a photographic plate, hence the term "spectrograph," whereas in the spectrometer the recording is performed electronically.

The mass spectrometer consists of the following essential parts: the high vacuum system, a system to introduce the sample, an ion source or ionization chamber, an electric field at high voltage, a magnetic field analyzer, a system to collect and record the ions. Let us briefly examine the various parts, beginning with the high vacuum system.

1. The mass spectrometer must operate under high vacuum. Therefore every instrument has a complex system of valves and pumps capable of reaching and maintaining the degree of vacuum required.

The pumps used are those of the oil or mercury diffusion type, but these must be connected to a preliminary pump, which is usually of a mechanical type. A diffusion pump holds the ion source at a vacuum of $10^{-7}/10^{-6}$ mm Hg., and another such pump maintains a vacuum of $10^{-8}/10^{-7}$ mm Hg. in the analyzer.

2. As for the system of introducing the sample, the gases and the volatile liquids can be kept in a reservoir from which they diffuse into the ionization chamber through a septum.

Samples of low volatility (for example, polar substances, compounds of high molecular weight, polymers, and many others) take advantage of a direct inlet system (DIS) into the ion source. Such a method gives a spectrum without resorting to excessive heating, which may decompose, dehydrate, or isomerize the compound. It is only necessary to operate at such a temperature that the pressure of the vaporized compound is in the order of $10^{-7}/10^{-6}$ mm Hg.

In the last few years the usefulness of the gas chromatograph as the inlet system has been shown in the analyses of complex mixtures of volatile substances. The separated substances flow through a molecular separator, connected to a vacuum system, and then into the ion source, where they become ionized. The molecular separator is present to remove most of the carrier gas, usually helium, and to allow the maximum possible yield of the resulting fractions to flow from the gas chromatograph into the ion source.

3. Let us consider now the ion source or ionization chamber, which is the "heart" of the instrument.

The substance can be ionized by various methods. The most widely used method is that of electron impact. Other types of ion source employed are field ionization, chemical ionization, surface ionization, spark ionization, field desorption, and photoionization.

By the electron bombardment method, the molecules of the sample, in a gaseous phase, enter the ion source through a slit. In the ion source a beam of electrons is generated by a rhenium or tungsten filament electrically heated to 2,000°C.

The energy of the electrons that bombard the substance is usually in the region of 8–100 eV. The effect of a collision between the electrons and the molecules of the sample is to generate positive ions (and, rarely, negative ions).

4. The ions become accelerated in the electric field, which has a potential difference of some thousands of volts (800–8000 V).

These ions pass through a slit of variable aperture (0.2–0.02 mm) and then enter the magnetic field analyzer.

5. Variation of the magnetic field, from a minimum to a maximum value of 20,000 gauss, gives each ion a curved trajectory, the radius of which depends on the $^{m}/e$ ratio for that ion. By this method, all ions having the same $^{m}/e$ value are combined in an ionic beam, and every beam, traveling along trajectories of diverse radii of curvature, ends its journey by passing through another slit and impinging on the collector, where it generates a signal the intensity of which is proportional to the number of ions arriving.

6. To record the ions a photographic or an electronic method can be used. In the photographic method an emulsion on a plate becomes impressed with the points of impact of single ionic beams. This type of recorder is employed in the mass spectrograph.

In the electronic system, the ionic beam that arrives at the collector produces an electric current the intensity of which is proportional to the number of ions contained in each beam. This current can be measured accurately and with great sensitivity by a type of Faraday cage or by an electron multiplier. The latter can register a current of 10^{-8} Amp. Thus it is possible to record a single ion arriving at the collector and still obtain a very clear mass spectral peak. The record of these signals constitutes the mass spectrum, from which the $^m/e$ values and the relative abundances can be deduced. In general the mass spectrum is recorded on light-sensitive paper.

Figure 2 shows a scheme of a mass spectrometer coupled to a gas chromatograph.

C. FUNDAMENTAL EQUATION OF MASS SPECTROMETRY.

To derive the fundamental equation of mass spectrometry, consider the electric and magnetic fields of a mass spectrometer of the magnetic deflection type.

In the electric field the potential energy (eV) of an ion is equal to its kinetic energy ($\frac{1}{2}mv^2$) after complete acceleration.

$$eV = \tfrac{1}{2}\, mv^2 \tag{1}$$

where e = electric charge,
V = acceleration potential of an ion,
m = mass of an ion,
v = velocity of an ion.

In the magnetic field the ion is subject to two forces in equilibrium, the centripetal force (Hev) and the centrifugal force (mv^2/R).

$$Hev = \frac{mv^2}{R} \tag{2}$$

where e, v, and m are as in (1),
H = intensity of the magnetic field,
R = radius of curvature.

Rearranging equation (2) and substituting v for the value in equation (1) we have

$$\frac{m}{e} = \frac{H^2R^2}{2V} \tag{3}$$

This is the fundamental equation of mass spectrometers of the magnetic deflection type.

From this equation we can see that a mass spectrometer can be maintained at constant H, with V varied (electric scanning), or V

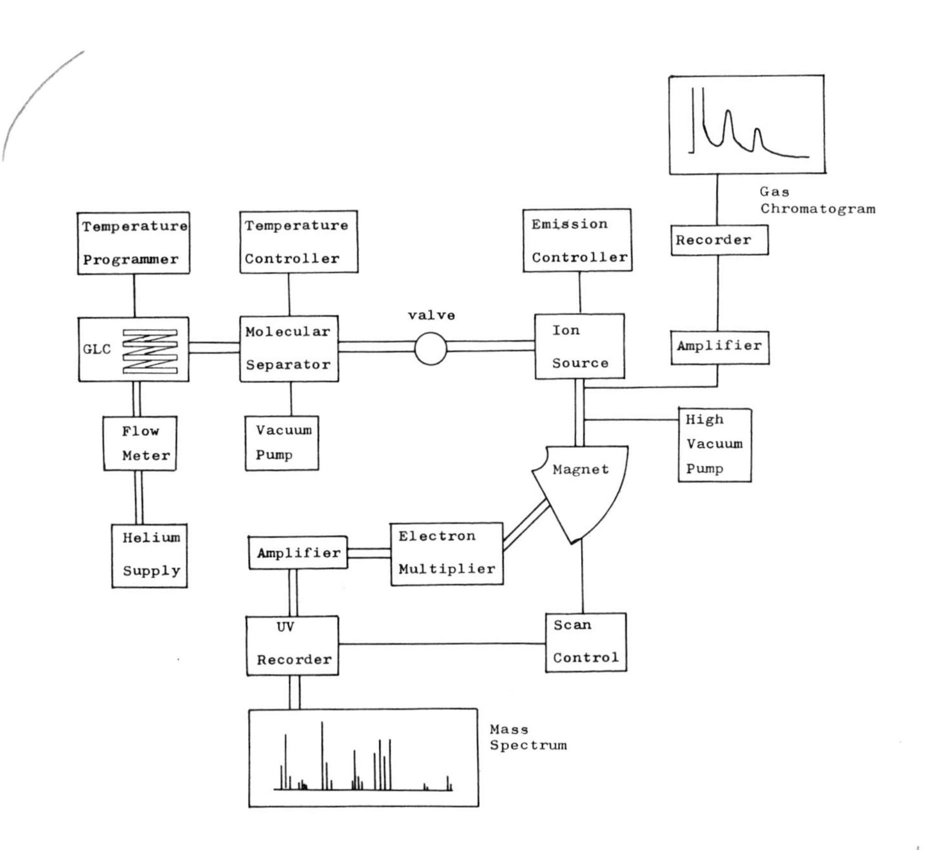

Figure 2 Scheme of a mass spectrometer coupled to a gas chromatograph.

next 93

can be maintained and H varied (magnetic scanning). The latter is the usual mode of operation.

Other mass spectrometers in addition to magnetic ones, such as the quadrupole, time-of-flight, and cycloidal mass spectrometers, are also available.

D. THE MASS SPECTRUM.

Figure 3 is an example of the mass spectrum of a substance, bufotenine, recorded on photosensitive paper, by a series of three galvanometers.

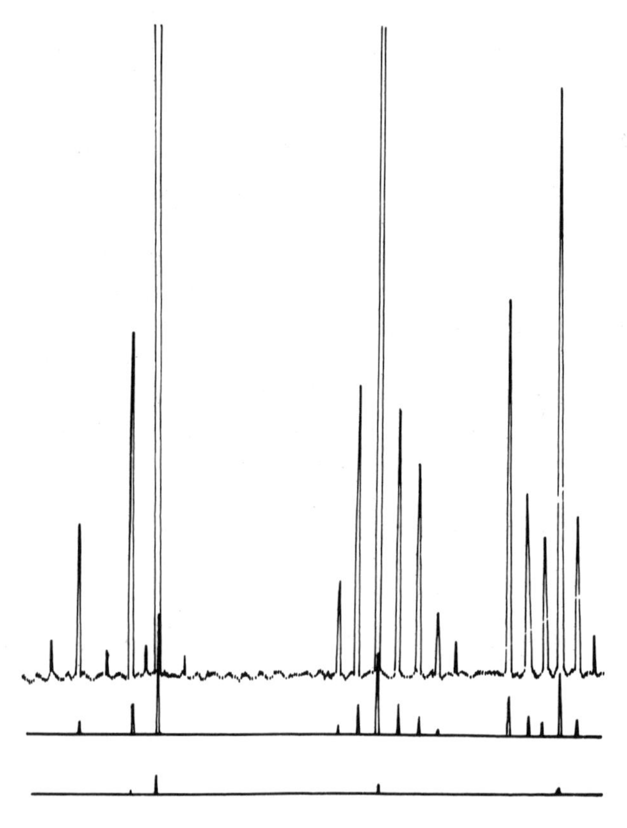

Figure 3 Partial mass spectrum of bufotenine as obtained from the instrument.

This method of recording helps us to see the peaks of very low relative abundance, which may be very significant.

Normally the whole spectrum is examined, from the lowest mass to the highest. It is also possible to examine a very restricted section of the spectrum. The process of assigning the peaks corresponding to each ion is called "counting the mass spectrum." Usually the count begins from mass 18 ($H_2O^{+\cdot}$), 28 ($N_2^{+\cdot}$), 32 ($O_2^{+\cdot}$), or 44 ($CO_2^{+\cdot}$), and because the distance between the peaks decreases logarithmically with increasing mass number, an interpolation may be necessary for the peaks at higher mass numbers.

Most modern instruments have an artificial means of numbering the masses, the "mass marker." This is an electronic instrument that provides a scale under the mass spectrum with a notch for every mass number.

The mass spectrum obtained from the spectrometer is rather difficult to study, and it is not possible to have a complete and immediate vision of what has happened.

The data can be presented in tabular form (figure 4), listing the $^m/e$ values and the corresponding relative abundances of every peak. The height of a peak is expressed as a percentage of the highest peak in the spectrum, called the "base peak," to which the arbitrary intensity of 100 is assigned. The tables are very practical for quantitative analysis but are not suitable for the interpretation of the mass spectrum of a given unknown substance because they do not show, in diagrammatic form, the important characteristics of the spectrum. In such cases it is

m/e	41	42	43	56	57	58	59	60	63	64	65	66	76
Rel. ab.	5	30	10	5	12	100	40	1	7	1	10	2	3
	77	78	88	89	90	91	92	101	102	103	104	105	115
	12	4	1	9	5	15	2	2	4	6	3	5	2
	116	117	118	119	128	129	130	131	132	133	144	145	146
	4	9	5	2	5	1	12	5	3	2	2	5	37
	147	148	157	158	159	160	161	201	202	203	204	205	
	6	1	1	5	13	13	3	2	3	2	44	7	

Figure 4 Tabular form of the mass spectrum of bufotenine.

preferable to represent the mass spectrum by a "line" spectrum on graph paper (bar-graph), where the abscissa is graduated in $^m/e$ values and the ordinate gives the percentage relative abundance to the base peak of the spectrum (intensity 100), as shown in figure 5.

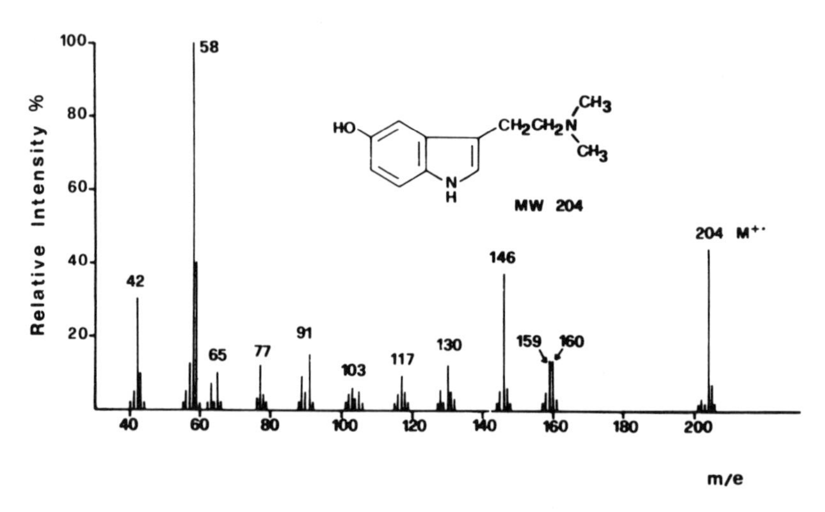

Figure 5 Normalized spectrum of bufotenine.

E. HIGH AND LOW RESOLUTION.

In a low resolution mass spectrometer, the atomic weight of the most abundant isotopes of the chemical elements are approximately whole numbers, for example,

$^{1}H = 1$	$^{16}O = 16$
$^{12}C = 12$	$^{19}F = 19$
$^{14}N = 14$	$^{32}S = 32$

From a physical point of view, assuming the most abundant isotope of carbon to be equal to 12 exactly, the other atoms are not equal to whole numbers, and their actual atomic weights can be determined to the sixth decimal place. On this precise scale of values, the atomic weights of some elements are as follows (table 1):

Table 1

^{1}H	=	1.007825
^{12}C	=	12.000000
^{14}N	=	14.003074
^{16}O	=	15.994915
^{19}F	=	18.998405
^{32}S	=	31.972074

From these values it can be seen that several combinations of atoms with the same whole number show a significant difference in their actual masses. Take for example, mass 28, common to carbon monoxide, a nitrogen molecule, and ethylene.

$$CO \ ^{m}/e = 27.994914$$
$$N_2 \ ^{m}/e = 28.006158$$
$$C_2H_4 \ ^{m}/e = 28.031299$$

To separate these ions from one another and to measure their masses accurately requires good resolving power.

F. RESOLVING POWER.

The "resolving power" of a mass spectrometer is defined as the capacity to separate ions of consecutive mass number. The resolving power necessary to separate two ions of mass m_1 and m_2 respectively is expressed by the following relationship.

$$RP = \frac{m_1}{m_2 - m_1} = \frac{m_1}{\Delta m} \qquad (4)$$

where $m_1 < m_2$

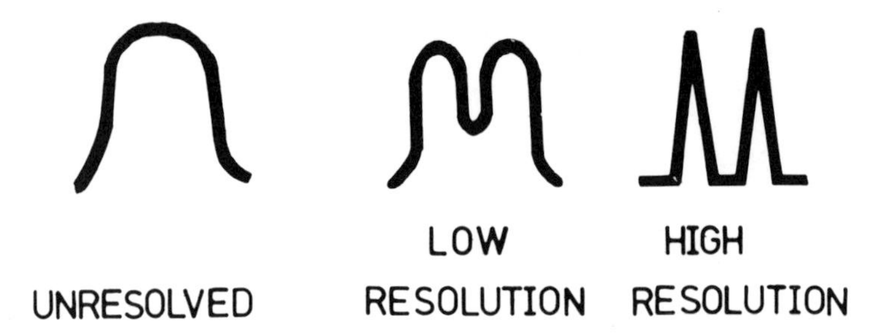

Figure 6 Diagram showing three states of resolution.

To clarify the significance of resolution let's look at figure 6. On the left is the case of "unresolved peaks," whereas on the right is that of "high resolution." There is no absolute rule for the resolving power of an instrument because that power depends on the convention adopted to determine when two peaks are considered to be resolved. In practice, two peaks are considered resolved when the "valley" between them is lower than a given percentage of the height of the two peaks (central diagram of figure 6). The most frequently used definitions are 10% or 50% valleys. In the 10% valley, two adjacent peaks of equal intensity have a valley 10% of the height of the peak. In the 50% valley, two adjacent peaks of equal intensity have a valley 50% of the height of the peak.

The upper limit of an instrument's resolving power is determined by such factors as the ion source, which is unable to produce an ionic beam completely homogenous in energy even under the best conditions; the magnetic field, which can focus a divergent ionic beam only if the ions all have the same energy; the aperture of the slits at the exit of the ion source and at the entrance to the collector; and the radius of the magnet.

What would be the resolving power of a mass spectrometer capable of separating the peaks of nitrogen and ethylene?

$$RP = \frac{m_1}{\Delta m} = \frac{28.006158}{28.031299 - 28.006158} = \frac{28.006158}{0.025141} = 1,110$$

Only a mass spectrometer with a resolving power greater than 1,110 would be able to separate the peaks.

Similarly, to separate the nitrogen and carbon dioxide peaks,

$$\mathrm{RP} = \frac{27.994914}{28.006158 - 27.994914} = \frac{27.994914}{0.011244} = 2{,}490$$

G. SINGLE FOCUSING MASS SPECTROMETERS.

The principal method, still employed in some mass spectrometers based on Dempster's 1918 instrument, is the deflection of an ionic beam through 180° in a magnetic field. The ions are initially accelerated by an electric field. The scheme of the instrument is illustrated in figure 7.

The ions produced in the source *a* are accelerated by a variable potential difference applied to plates *b*. They enter the magnetic field H by slit *c*. After the beam has been deflected through 180° the ions are focused through slit *e* and impinge on the collector *f*. This instrument is defined as a "direction" or "single" focusing spectrometer.

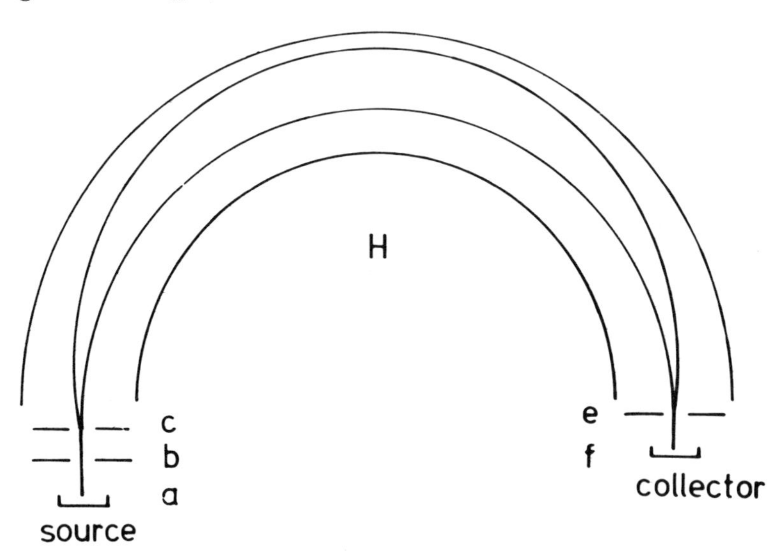

Figure 7 180° Magnetic deflection mass spectrometer.

The mass spectrum is obtained when the ions of different mass arrive at the collector, one after the other. The time of arrival will be proportional to the square root of the $^{m}/e$ ratio.

In more recent instruments the 180° magnetic field is substituted with one of the sector type, as shown in figure 8, where $\vartheta = \varphi = 30°$ or $45°$. The sector instruments have become competitive with those of the 180° type and have the advantage of using a much smaller magnet.

H. DOUBLE FOCUSING MASS SPECTROMETERS.

The double focusing instruments have been investigated as a means of achieving a resolving power of some thousands. It is important to eliminate the difference in kinetic energy between the ions of the same mass before the beam enters the magnetic analyzer. This is done by placing an electrostatic analyzer between the ion source and the magnetic field, as shown in figure 9.

The radius of curvature (R) of an ion in the electrostatic field is given by

$$R = \frac{2V}{E} \qquad (5)$$

where V is the potential at which the ion is accelerated out of the source and E is the potential applied to the electrostatic analyzer plates.

It is evident therefore that the trajectory of the ions is not only a function of the $^{m}/e$ ratio but also of the kinetic energy.

The electrostatic analyzer consists of a tube of fixed radius under a constant electric field. Only ions the kinetic energy of which corresponds to the radius of curvature are able to pass into the magnetic analyzer. By this method there is a "focusing of velocity" before the directional focalization.

This combination is called "double focusing." It eliminates the principal causes of low resolution and produces a coherent ionic beam with respect to the kinetic energy.

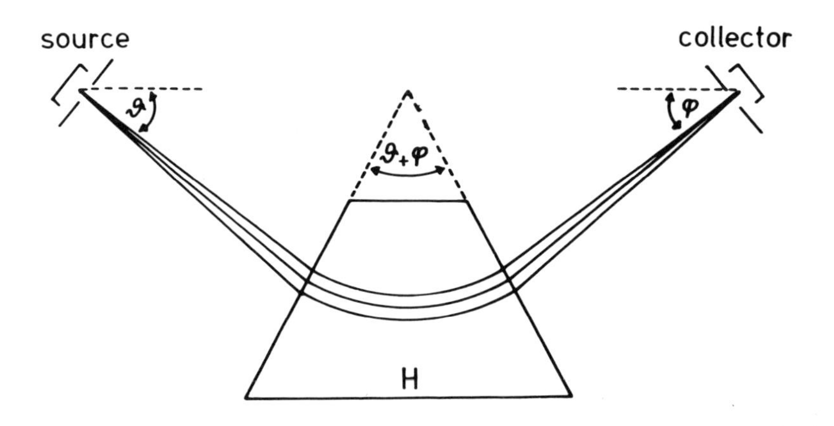

Figure 8 Sector type of magnetic deflection mass spectrometer.

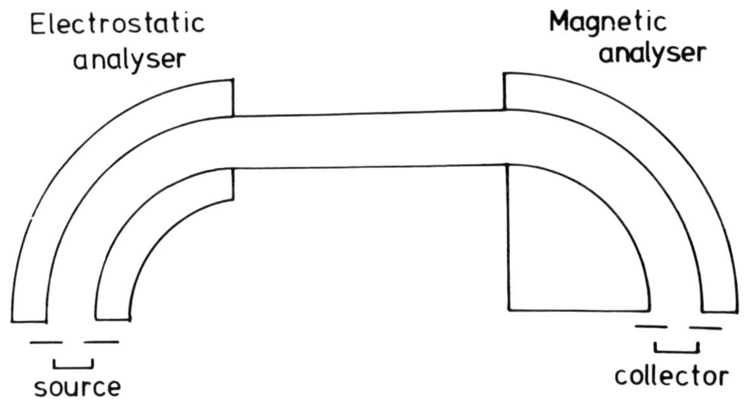

Figure 9 Double focusing mass spectrometer.

The most widely used optics in this type of mass spectrometer are the Mattauch-Herzog and Nier-Johnson types.

The Mattauch-Herzog system consists of a combination of a radial electrostatic analyzer and a magnetic analyzer, in series, that focus the ions onto a plane where there is a photographic plate. The ions produced in the source pass through an electric field, a 31°50′ sector, and then through a 90° magnetic sector (figure 10). With instruments of this type, following two successive focalizations, the resolving power can be in the order of several thousands. These optics are applicable to spectrometers, but are more usually applied to spectrographs.

The Nier-Johnson optics consist of a combination of a 90° electrostatic sector and a 60° or 90° magnetic sector, in series, as shown in figure 11.

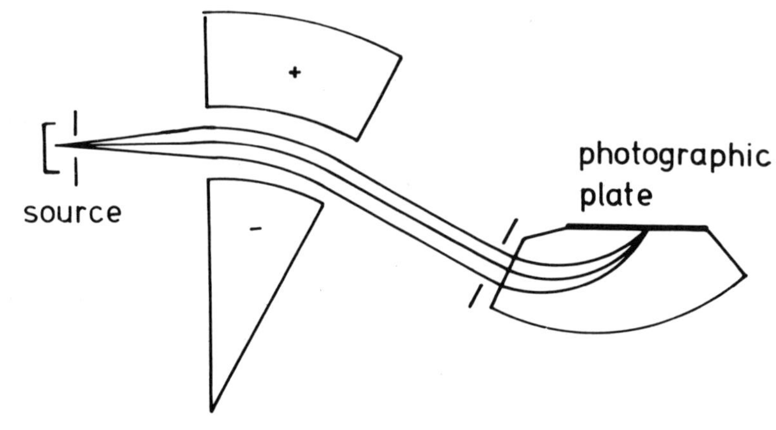

Figure 10 Mattauch-Herzog optics.

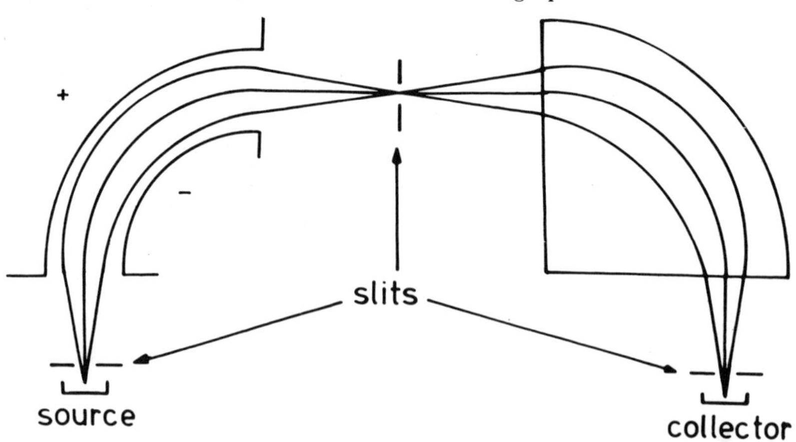

Figure 11 Nier-Johnson optics.

This type is differentiated from the former by the fact that the ionic beams, having left the electrostatic analyzer, are convergent, and it is therefore possible to put another slit between the electrostatic and magnetic analyzers.

By varying the intensity of the magnetic field the separated ionic beams reach the slit of the collector one after the other. At the collector they are recorded. This system is only applicable to a spectrometer, *not* to a spectrograph. The resolving power is comparable with that of the Mattauch-Herzog optics.

I. THE QUADRUPOLE MASS SPECTROMETER.

The operative principle of the quadrupole mass spectrometer is given by the Mathieu equations which describe the trajectory of a particle moving through the lines of force of two fields at continuous current modulated by a radiofrequency and applied to a quadrupole system.

In practice, the analyzer of this instrument is made up of four metal bars, fixed at the angles of a square and ceramically isolated, connected alternately to form two couples to which are applied DC and RF potentials with charges of opposite sign, as illustrated in figure 12.

The accelerated particle enters the analyzer (filter) and begins to oscillate in a complex manner according to the $^{m}/e$ and RF/DC ratios. Some typical trajectories are shown in figure 13. For every value of these ratios only one mass is able to pass completely through the filter and impinge on the collector. The others, having trajectories of greater periodicity, impinge on one of the charged poles and are discharged. Therefore, by varying the $^{RF}/DC$ ratio it is possible to select a single value of $^{m}/e$ able to give a signal, and by continuous variation, from a minimum to a maximum, the whole spectrum can be recorded. One peculiarity of the quadrupole mass spectrometer is that it supplies linear spectra—that is, the distance between two successive mass units is constant.

The resolving power of this instrument is about 1,000.

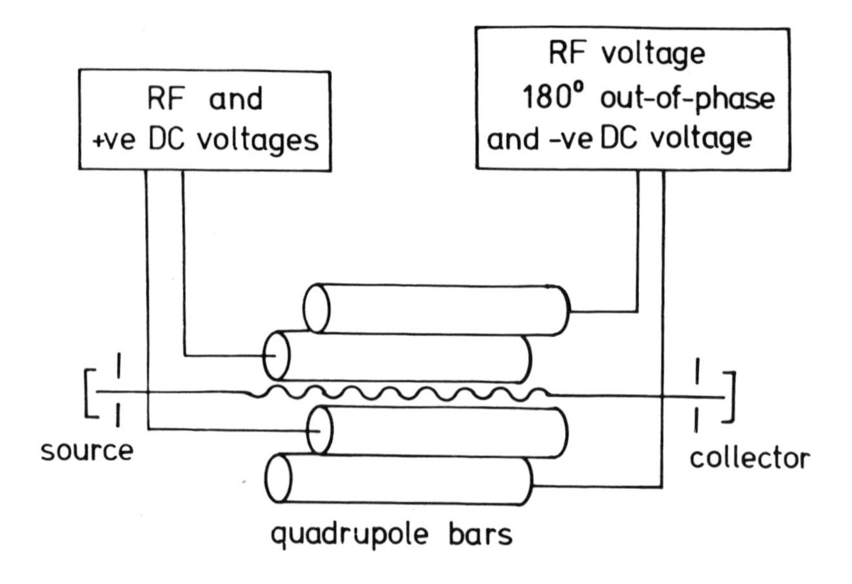

Figure 12 Quadrupole mass spectrometer.

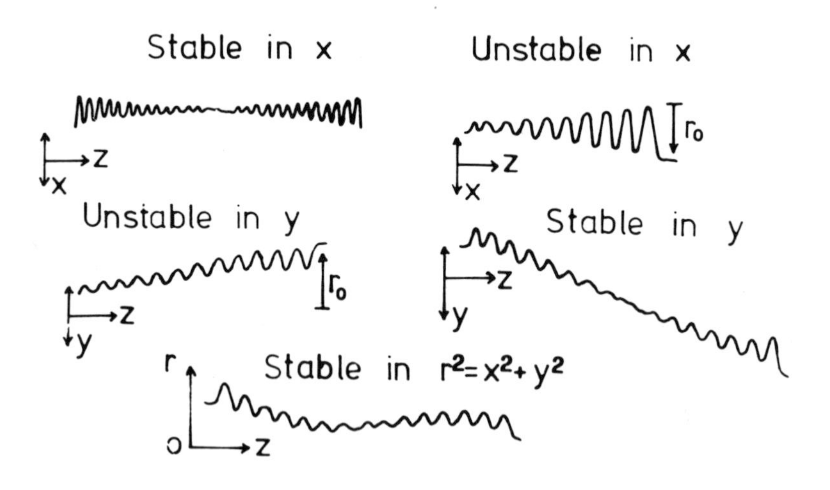

Figure 13 Various stable and unstable trajectories.

J. DIAGRAMMATIC SUMMARY.

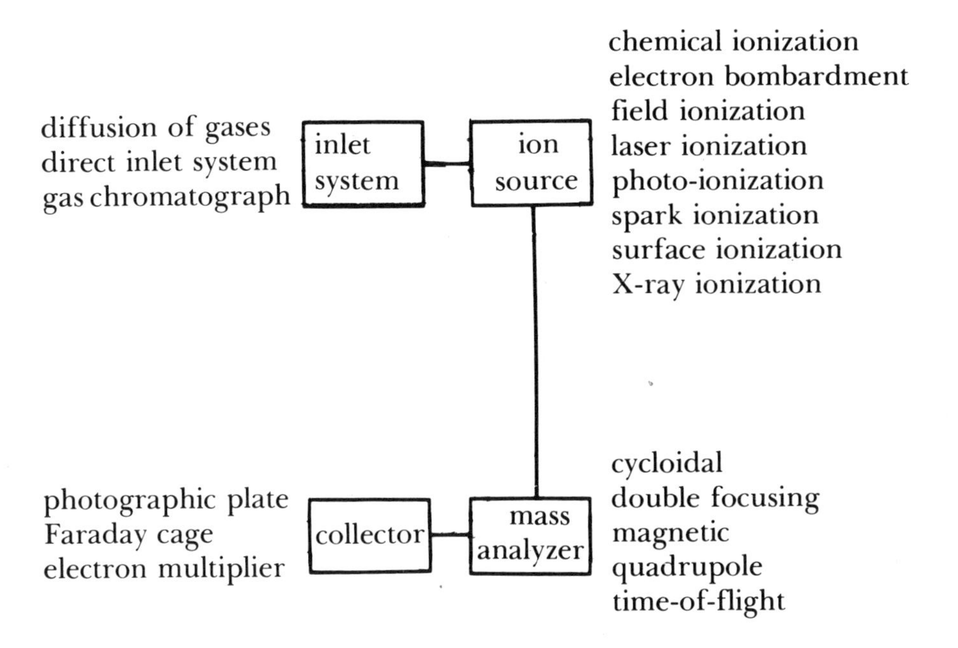

Figure 14 Scheme showing the different combinations for the mass spectrometer.

CHAPTER II

A. THE PRINCIPAL TYPES OF IONS.

The ions that appear in the mass spectrum are molecular ions, isotopic ions, fragment ions, rearrangement ions, multiply charged ions, metastable ions, negative ions, and ions by ion-molecule interaction.

1. Molecular Ions.

The ion formed by the loss of one electron from part of the molecule is called the "molecular ion" or the "parent ion." The corresponding peak in the mass spectrum is called the "molecular peak" or the "parent peak."

The molecular ion is the precursor of all other ions in the spectrum and requires the minimum energy for its formation. The nominal mass of such an ion corresponds to the molecular weight of the compound. It is important to note that the mass number obtained by this method is the exact molecular weight and not an approximate value as obtained by classical chemical methods such as cryoscopy and ebullioscopy. The relative intensity of the molecular ion depends on its stability with respect to the decomposition products, and it thus indicates the type of compound under examination. For example, the aromatic compounds give rise to an abundant molecular peak because of the π electrons in the system. Cyclic molecules also have stability because of the fact that at least two bonds must be broken before they fragment. In general, every structural characteristic that favors the delocalization of the electronic charge leads to an increase in stability and thus to a particularly intense peak. In figure 15 two examples of extremely simple molecular ions are shown.

$$CH_4 + e^- \longrightarrow CH_4^{+\bullet} + 2e^-$$

$^m/e$ 16

$$CH_3{-}CO{-}CH_3 + e^- \longrightarrow CH_3{-}CO{-}CH_3]^{+\bullet} + 2e^-$$

$^m/e$ 58

Figure 15 Formation of two extremely simple molecular ions.

2. Isotopic Ions.

The molecular ion contains the most abundant isotopes of the constituent elements. The parent peak is always accompanied by other peaks of higher mass that are caused by the ions which contain the heavier isotopes.

For example, CH_3I has a molecular ion at $^m/e$ 142, derived from the lightest isotopes of the constituent elements $^{12}C^1H_3{}^{127}I$, and there is a peak at $^m/e$ 143 due to $^{13}C^1H_3{}^{127}I$. This peak, 143, is 1.1% of the 142 peak because the natural abundance of ^{13}C is 1.1% of ^{12}C. Benzene shows an intense peak at $^m/e$ 78 for the ion $^{12}C_6{}^1H_6$ and another peak at $^m/e$ 79 due to the ion $^{13}C^{12}C_5{}^1H_6$ and, in part, to the ion $^{12}C_6{}^1H_5{}^2H$. Neglecting the isotopic contribution of the latter ion, since ^2H is present in only 0.015% in nature, the intensity of peak 79 is 6.6% of peak 78 because every atom of carbon has a 1.1% probability of being ^{13}C and therefore $1.1\% \times 6 = 6.6\%$.

Inversely one can infer that a peak of mass 142 that is accompanied by a peak of mass 143 and with an intensity of 1.1% with respect to 142 cannot contain more than 1 atom of carbon. One can also infer that an ion of mass 72 that has a peak 73 with a relative intensity of 4.5% cannot include more than 4 carbon atoms and must have a heteroatom such as oxygen or nitrogen in the molecule.

The isotopic abundances (%) of the most frequently occurring elements are as shown in table 2.

Table 2

^{1}H = 99.985	^{2}H = 0.015		
^{12}C = 98.89	^{13}C = 1.108		
^{14}N = 99.64	^{15}N = 0.36		
^{16}O = 99.76	^{17}O = 0.04	^{18}O = 0.20	
^{32}S = 95.06	^{33}S = 0.66	^{34}S = 4.20	^{36}S = 0.14
^{35}Cl = 75.4	^{37}Cl = 24.6		
^{79}Br = 50.57	^{81}Br = 49.43		

The mass spectrometer is particularly useful for demonstrating the presence of chlorine, bromine, and sulphur atoms. These atoms have abundant isotopes (especially Cl and Br) two mass units higher, and they give peaks with characteristic intensities at masses m, m+2, m+4, etc. The interval is two mass units because the isotopes of the elements differ by two neutrons.

Let us take a monochloride as an example. As we have seen, chlorine consists of isotopes ^{35}Cl (75.4% of all chlorine) and ^{37}Cl (24.6% of all chlorine), and these two isotopes are in the ratio 3:1 approximately. In the mass spectrum the chlorides are recognized by their isotopic distribution, the two peaks being two mass units apart and with an intensity ratio of 3:1 between mass m (from ^{35}Cl) and mass m+2 (from ^{37}Cl).

In general, the contribution of the various isotopes of the different elements can be calculated from the following expression.

$$(a+b)^n \tag{1}$$

where a = % natural abundance of the lighter isotope,
b = % natural abundance of the heavier isotope,
n = number of atoms of the element present in the molecule.

In the case of a molecule containing two atoms of chlorine, such as methylene chloride, CH_2Cl_2,

$$(a+b)^n = (0.754 + 0.246)^2 = 0.568 + 0.371 + 0.060 \equiv 100 : 65.25 : 10.60$$

The three ions—mass m ($CH_2{}^{35}Cl_2$), m+2 ($CH_2{}^{35}Cl^{37}Cl$), and m+4 ($CH_2{}^{37}Cl_2$)—have intensity ratios of 100: 65.25 : 10.60 respectively.

When a molecule is formed with two different poly-isotopic elements (for example Cl and Br) the following equation must be used,

$$(a+b)^n(c+d)^r$$

where a, b, c, and d refer to the percentage abundance of ^{35}Cl, ^{37}Cl, ^{79}Br, and ^{81}Br respectively, and n and r are the number of atoms of chlorine and bromine respectively.

Tables exist in which the relative abundance of the various isotopes are given.

3. Fragment Ions.

The fragment ions are formed by the fragmentation of the molecular ion in the ion source. Because of the considerable energy of the electronic bombardment usually employed to ionize an organic substance (70 eV), the molecular ion, being in an excited state, tends to break some of its interatomic bonds, producing fragments of lower mass called "ionic fragments."

The study of the fragmentation patterns of the most abundant ions can furnish structural information on the compound analyzed (figure 16).

$$CH_4 + e^- \longrightarrow CH_4^{+\cdot} + 2e^-$$
$$\downarrow$$
$$CH_3^+ + H^\cdot$$

$CH_4^{+\cdot}$ $^m/e$ 16 (molecular ion), CH_3^+ $^m/e$ 15 (fragment ion)

$$CH_3-CO-CH_3 + e^- \longrightarrow CH_3-CO-CH_3^{+\cdot} + 2e^-$$
$$\downarrow$$
$$CH_3-C\equiv O^+ + CH_3^\cdot$$

Figure 16 Formation of two extremely simple fragment ions.

4. Rearrangement Ions.

The ions formed by rearrangement or by transposition are those whose origin cannot be explained by the simple rupture of a bond in the molecule. The classical and most common example is called the "McLafferty rearrangement." This mechanism is confined to compounds such as aldehydes, ketones, acids, and esters, which contain a suitable hydrogen atom on the γ-carbon atom with respect to the carbonyl group. Such a process involves a six-membered cyclic transition state (figure 17).

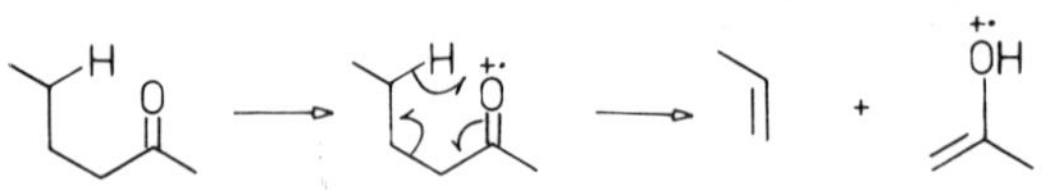

Figure 17 Typical six-membered cyclic transition state.

5. Multiply Charged Ions.

The ions that are normally formed have a single charge because they have lost one electron from the molecule and consequently formed a molecular ion in a highly excited state. With successive fission the molecular ion forms ionic fragments of single charge. Some molecules are very stable and can sustain the loss of two or more electrons, giving rise to ions at $^{m}/2e$ and $^{m}/3e$ respectively. They are recorded at ½ or ⅓ the mass of a singly charged ion. The aromatics and molecules containing a conjugated system can stabilize an ion with a double charge because of the presence of the π electron system. The presence of a multiple charge indicates that an unusually stable molecule is under analysis.

6. Metastable Ions.

The presence of metastable ions is indicated in the mass spectrum by broad peaks, called "metastable peaks," that do not usually appear at whole mass numbers. Metastable peaks are useful in the study of fragmentation patterns because they demonstrate that there is a correlation between certain ions formed by a unimolecular decomposition.

Let us consider the average lifetime of an ion m_1 formed in the ion source.

1. If the average lifetime of the ion is $\gg 5 \times 10^{-6}$ secs. the ion becomes accelerated, deviated, and recorded as m_1.

2. If the average lifetime of ion m_1 is $\ll 5 \times 10^{-6}$ secs. it fragments before acceleration to give a new ion of mass m_2,

$$m_1^+ \longrightarrow m_2^+ \quad + \quad (m_1 - m_2) \tag{2}$$

where $(m_1 - m_2)$ represents a neutral fragment. Only m_2 will be recorded on the mass spectrum.

3. If the average lifetime is such that the ion is decomposed in the zone between the source and the analyzer, there will be an ion with the acceleration of m_1 (velocity v_1) that becomes deviated like m_2 (still with velocity v_1). The ion m_2 appears in the spectrum at m*, which is given by the relationship,

$$m^* = \frac{{m_2}^2}{m_1} \tag{3}$$

Such a decomposition is called a "metastable transition," and the peak m* is the metastable peak. The ions correlated from a metastable peak indicate that a neutral fragment has been lost in a single process. If one hypothesizes a fragmentation reaction, the hypothesis can be confirmed by the presence of metastable peaks. Figure 18 is an example of a metastable transition.

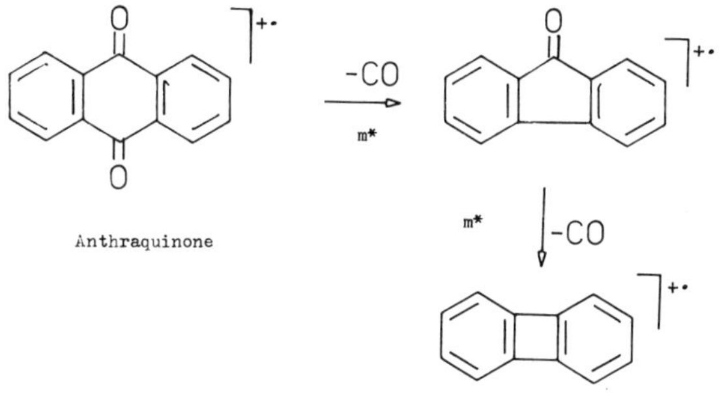

Figure 18 Example of a metastable transition.

7. Negative Ions.

Negative ions formed in the ion source by electron bombardment are less abundant than positive ions by a factor of 10^{-4}. The study of negative ions has made little progress and constitutes an open field for future research. These ions are formed from a neutral molecule by three mechanisms.

1. production of a couple of ions,

$$AB + e^- \longrightarrow A^+ + B^- + e^-$$

2. capture of an electron, with dissociation,

$$AB + e^- \longrightarrow A + B^-$$

3. capture of an electron,

$$AB + e^- \longrightarrow AB^-$$

8. Ions by Ion-Molecule Interactions.

The percentage of molecules of a compound ionized to the vapor in the ionization chamber is very low. Therefore it is probable that a collision will occur between a molecular ion and a neutral molecule. In such a collision the molecular ion can subtract an atom or group from the neutral molecule, forming a heavier ion.

The abundance of these ions is proportional to the square of the pressure of the sample because they are formed by a second-order process. In general the pressure used is very low for a normal analysis, and so these ions have negligible abundance. They appear in significant abundance if the molecular ion is unstable and the corresponding protonated molecular ion is highly stable. This situation gives rise to a very small or completely negligible M peak but a substantial M+1 peak.

This behavior is found in the spectra of some ethers, esters, amines, aminoesters, and nitriles. The M+1 peak is useful for determining the molecular weight in these circumstances.

B. INCREASING THE VOLATILITY OF A SUBSTANCE.

One of the major limitations of mass spectrometry is the fact that it requires a vaporized sample. Chemical pretreatment can be used to prepare a derivative that is either more volatile or has a characteristic fragmentation pattern. The derivatizations must be applicable to a very small quantity of product, often 1 mg of sample. Some extremely useful derivatives are listed in table 3.

Table 3

Compound	Derivative
ROH	$ROCOCH_3$
ROH	$ROSi(CH_3)_3$
RCOOH	$RCOOCH_3$
$RCONH_2$	RCH_2NH_2

A knowledge of the factors that govern the fragmentation of a molecular ion is an obvious advantage when deciding which derivative is the most suitable for a mass spectrometric examination of the sample.

C. FORMATION OF THE MOLECULAR ION.

When a molecule of the sample, in the vapor phase in the high vacuum of the instrument (10^{-6} mm Hg), is bombarded with electrons of the correct energy, there are two types of process it can undergo: (a) the extraction of one or more electrons (reactions 4 and 5), or (b) the absorption of an electron by the molecule (reaction 6).

$$ABCD + e^- \longrightarrow ABCD^{+\cdot} + 2e^- \qquad (4)$$
$$ABCD + e^- \longrightarrow ABCD^{n+} + (n+1)e^- \qquad (5)$$
$$ABCD + e^- \longrightarrow ABCD^- \qquad (6)$$

The ion $ABCD^{+\cdot}$ is called the molecular ion and is indicated by $M^{+\cdot}$.

The negative ions, which are formed in negligible quantities compared with the positive ions, do not enter the magnetic field because they are not accelerated out of the ionization chamber. They can therefore be ignored.

The formation of a positive ion occurs when the energy of the bombarding electron is at least equal to the ionization potential of the molecule, which varies from 7 V to 15 V (for benzene 9.2 V, for methane 14.0 V). If the energy of the bombarding electrons equals the ionization potential of the molecule, there must be a complete transfer of energy from the electron to the molecule for the ionization to occur. At this energy level the abundance of the molecular ion is very small.

An increase in the energy of the electrons increases the probability of ionization and therefore increases the abundance of the molecular ion.

A graph of the relative intensity of the molecular ion against the energy of electron bombardment is called "the efficiency curve of ionization." An example is shown, for pyridine, in figure 19.

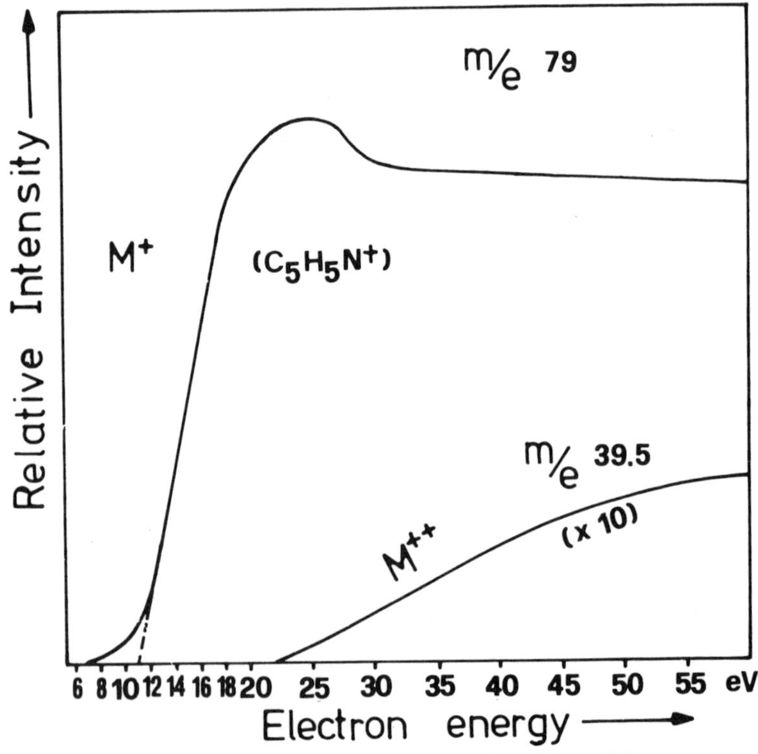

Figure 19 Efficiency curve of ionization for pyridine.

The efficiency curve rises sharply when the energy of the electron coincides with that of ionization (dotted lines), it reaches a maximum at 20–25 eV, and it levels out at about 30–50 eV.

The electrons produced in the ion source are not rigorously monoenergetic; rather the energy is distributed about a Gaussian curve in an interval of ± 1 eV about an average value. There are always some electrons of higher energy with respect to the average, and, therefore, in reality the curve does not rise sharply but assumes a form indicated by the continuous line (figure 19).

An increase in the energy of the electrons can cause a double ionization of the molecule to occur. The excess energy of the molecular ion reaches a value equal to or greater than the energy required to break the bonds of the molecule, and so the molecular ion fragments. A graph of the abundance of the doubly charged molecular ion and the fragments, as a function of the energy, shows a curve similar to the efficiency curve for ionization of the molecule (figure 19).

All the curves level out at energies higher than 40–50 eV. Therefore, to obtain a reproducible mass spectrum, an instrument with an electron bombarding energy of 50–80 eV is used.

It is important to make a distinction between the ionization potential and the appearance potential. The ionization potential is the minimum electronic energy required to produce a molecular ion, as in reaction (4).

For the molecular ion, the appearance potential coincides with the ionization potential; whereas, for a fragment ion, the appearance potential is the minimum energy required to produce the fragment ion from a molecular ion.

For a molecule R_1R_2 that ionizes and fragments as in reaction (7),

$$R_1R_2 + e^- \longrightarrow R_1R_2^{+\cdot} + 2e^- \qquad (7)$$
$$R_1R_2^{+\cdot} \longrightarrow R_1^+ + R_2^{\cdot}$$

the appearance potential of the ion R_1^+—that is, $A(R_1^+)$—is given by equation (8),

$$A(R_1^+) = D(R_1-R_2)^+ + I(R_1R_2) + E \qquad (8)$$

where $D(R_1-R_2)^+$ is the dissociation energy of the R_1-R_2 bond, $I(R_1R_2)$ is the ionization potential of molecule R_1R_2, and E is the excess vibrational, electronic, and kinetic energies of fragments R_1^+ and R_2^+.

Usually E is small and negligible. The accurate determination of I and A is not very easy because of the dispersion in energy of the electron beam emitted from the hot filament (the energies are compressed into an interval of ± 1 eV). Despite this, it is easy enough to calculate, for example, the dissociation energy of a bond from the ionization potential and appearance potential data. The dissociation energies determined in this way do not disagree greatly from those found by other methods. The ionization potential and the appearance potential are often used in the study of fragmentation processes. Because the ionization potentials of organic compounds vary greatly in relation to their structural characteristics, it is sometimes possible to determine to which class a compound belongs. Table 4 gives the ionization potentials of some substituted benzenes.

Table 4

Compound	I(V)
Nitrobenzene	10.18
p-Nitrotoluene	9.82
Benzene	9.56
Nitrophenol	9.52
Toluene	9.18
Phenol	9.16
p-Cresol	8.97
Anisole	8.83
Aniline	8.32
p-Methoxyaniline	7.82

Moreover, from such data the source of many fragment ions can be inferred. One of the most important applications of this method was the identification of the $C_7H_7^+$ ion at $^m/e$ 91 as a tropylium ion, in the spectra of alkylbenzenes. It was thought, initially, that the $C_7H_7^+$ ion, the origin of which could be the molecular ion of toluene, had the benzylic structure. The difference between the calculated value for the benzylic C-H bond energy as studied by pyrolysis (77 kcal) and the appearance potential (95 kcal) led to the exclusion of the benzylic ion but supported the tropylium ion.

D. DECOMPOSITION OF THE MOLECULAR ION.

The ionization of a molecule produces a molecular ion in an energetically excited state without any change in the bond lengths. According to the theory of "quasi-equilibrium," the energy of excitation is distributed in all the ions and transformed into vibrational energy in the various bonds. Some ions (activated complexes) possess sufficient energy to break one or more bonds and will thus give rise to a molecular ion decomposition. The decomposition reaction can be of two types—simple fragmentation, or fragmentation with rearrangement.

The decomposition of molecule ABCD can be represented by the following scheme.

$$ABCD + e^- \longrightarrow ABCD^{+\cdot} + 2e^- \text{ (ionization)} \quad (9)$$

(a)

$$ABCD^{+\cdot} \xrightarrow{k_1} A^+ + BCD^\cdot \text{ or } A^\cdot + BCD^+ \quad (10)$$

$$ABCD^{+\cdot} \xrightarrow{k_2} AB^+ + CD^\cdot \text{ or } AB^\cdot + CD^+ \quad (11)$$

$$ABCD^{+\cdot} \xrightarrow{k_3} ABC^+ + D^\cdot \text{ or } ABC^\cdot + D^+ \quad (12)$$

$$AB^+ \xrightarrow{k_4} A^+ + B \text{ or } A + B^+ \quad (13)$$

etc. (Fragmentations)

(b)

$$ABCD^{+\cdot} \longrightarrow AD^{+\cdot} + BC \text{ or } AD + BC^{+\cdot} \quad (14)$$

(Fragmentation with rearrangement)

From a qualitative point of view it can be shown that the velocity of a particular decomposition reaction depends on the concentration of the activated complex and on the activation energy necessary to reach the transition state; that the relative abundance of the ions produced depends on the velocity of a single decomposition, that is, on the activation energy of the corresponding transition state; and that the abundance of a particular ion depends on the velocity of all the reactions that produce those ions and on the velocity of all their decomposition reactions.

Thus, the abundance of the ion AB^+ depends on k_2 and k_4 and the abundance of A^+ on k_1 and k_4.

Because the structure of the transition state is very similar to the product of the reaction, the activation energy of the reaction is usually influenced by the stability of the products formed. Consequently, a structural modification that gives rise to a product of higher energy content increases the energy of activation, whereas if it gives rise to a product of lower energy content the energy of activation is decreased. Also, a structural change that alters the lability of a bond of the reagent ion alters the activation energy. In a decomposition, neutral products, radicals, and ions are formed. In general it is the stability of the latter that is the determining factor.

CHAPTER III

A. DETERMINATION OF THE MOLECULAR WEIGHT.

The mass number corresponding to the peak of the molecular ion is the molecular weight of the substance under examination. Of all the methods employed until now for the determination of the molecular weight (such as cryoscopy, ebullioscopy, and osmometry), mass spectrometry is the only one that gives an exact figure; and it does so very quickly.

Because a knowledge of the exact molecular weight is important in the determination of the structure of a substance and for the correct interpretation of the mass spectrum, the characteristics of the molecular ion and the relative peak will be discussed in more detail, and this discussion will be followed by an exposition of the principal methods of establishing the molecular weight.

The stability of the molecular ion varies enormously from product to product. The various classes of organic compounds can be listed in order of decreasing stability of the molecular ion: aromatic compounds, conjugated olefins, alicyclics, sulphides, straightchain hydrocarbons, mercaptans, ketones, amines, esters, ethers, carboxylic acids, branched hydrocarbons, and alcohols.

In table 5 the intensities of the molecular ions are reported, expressed as a percentage of the total ion current, for some compounds of similar molecular dimensions.

Table 5

Substance	Abundance ($\%\Sigma_m$)
Naphthalene	44.3
Quinoline	39.6
n-Butylbenzene	8.26
trans-Decaline	8.22
t-Butylbenzene	7.00
Allo-ocimene	6.40
Diamylsulphide	3.70
n-Decane	1.41
n-Decylmercaptan	1.40
Diamylamine	1.14
Methylnonoate	1.10
Mircene	1.00
Cyclododecane	0.88
3-Nonanone	0.50
n-Decylamine	0.50
Diamyl ether	0.33
Cis-cis-2-Decalool	0.08
3-Nonalool	0.05
Linalool	0.04
3,3,5-Trimethyl-heptane	0.007
n-Decanol	0.002
Tetrahydrolinalool	0.000

In tetrahydrolinalool, the molecular ion is not present. The peak of highest mass number in its mass spectrum (M − 18) corresponds to the loss of water. If other methods, such as IR and NMR, indicate that the substance contains a hydroxyl group, the molecular weight can be calculated immediately.

Analogously, in the case of an acetate, in which the highest mass is $^m/e$ M−60, the molecular weight can be determined. Because of the high sensitivity of the mass spectrometer and the reproducibility of the spectra, a peak of 1% intensity is clearly visible and easily identifiable. The determination of the molecular weight does not present a problem if the substance is pure and the corresponding molecular peak has at least a 1% intensity. If, however, the molecular ion has a low intensity or is absent, the molecular weight can be determined indirectly. The formation of derivatives by a simple chemical reaction is a technique frequently used in mass spectrometry to obtain an intense molecular ion peak, to render the substance more volatile, and to acquire further information for the interpretation of the spectrum.

It has been stated previously that the formation of ionic fragments requires excess energy with respect to the ionization potential of the molecule. Therefore, by reducing the potential of electron bombardment, an increase in the abundance of the molecular ion can be obtained, expressed as a percentage of the sum of the intensities of all the ions. For example, in the spectrum of 3-methylindole, only two significant peaks exist, the molecular ion $M^{+\cdot}$ at $^m/e$ 131 and the M − ion. At 70eV, M − 1 is double the height of $M^{+\cdot}$, whereas at 12eV it is 20% of $M^{+\cdot}$, and at 9eV only $M^{+\cdot}$ is present.

The number of ions formed in the ionization chamber is small when compared to the number of molecules that are not ionized (the efficiency of ionization is in general one per million); thus a collision can occur between a positive ion and a neutral molecule.

During this collision the ion can subtract an atom or group from the neutral molecule and thus form an ion with a total mass greater than that of the molecule analyzed. The reaction can be written as in scheme (1),

$$\mathrm{ABCD}^{+\cdot} + \mathrm{ABCD} \longrightarrow \left[\ \underset{\displaystyle \mathrm{ABCDA}^{+} + \mathrm{BCD}^{\cdot}}{\underset{\downarrow}{\mathrm{ABCD.ABCD}}}\ \right]^{+\cdot} \qquad (1)$$

Because this is a bimolecular reaction, it depends on the concentration of the ions of the molecule and is proportional to the square of the concentration of the sample or to the square of its pressure. At the pressures usually employed in the mass spectrometer, the bimolecular reactions are negligible, except for the extraction of a hydrogen atom (to form the $(M+1)^+$ ion), when the molecular ion contains an atom of either oxygen or nitrogen.

When the pressure of the sample is raised slightly, the $(M+1)^+$ ion increases in intensity and becomes significantly diagnostic for those compounds in which the molecular ion is very unstable and has a great tendency to decompose, but the corresponding protonated molecule is stable (for example, oxonium and ammonium ions). This is the case for the alcohols, ethers, esters, amines, amino-esters, sulphones, carbonates, and nitriles in which the molecular weight can be determined easily in spite of the weak intensity of the molecular ion.

Recently new ionization methods have been introduced into the mass spectrometer that turn out to be useful in the study of the molecular ion. These are "field ionization" and "chemical ionization."

In "field ionization" the molecules of the sample are placed in the vicinity of a very thin wire at high potential (10–20 kV). Near the surface of the wire there is a field gradient on the order of 10^8 volts/cm, which is capable of ionizing the molecule. The ions formed are devoid of vibrational energy, and their tendency to dissociate is low compared with that of ions produced by electron bombardment. This procedure leads to an increase in the abundance of the molecular ion, which for many compounds is the only important ion in the spectrum. Figure 20 shows the spectra of 3,3-dimethyl-pentane as displayed by (a) electron bombardment and (b) field ionization.

The fragments from the two processes of ionization are qualitatively but not quantitatively the same, and in the case of field ionization the molecular ion is particularly intense.

In the case of "chemical ionization" the molecules of the compound under examination are introduced into the ion source of the instrument "diluted" with a large excess of a gas (usually methane, but also propane, iso-butane, hydrogen, or ammonia).

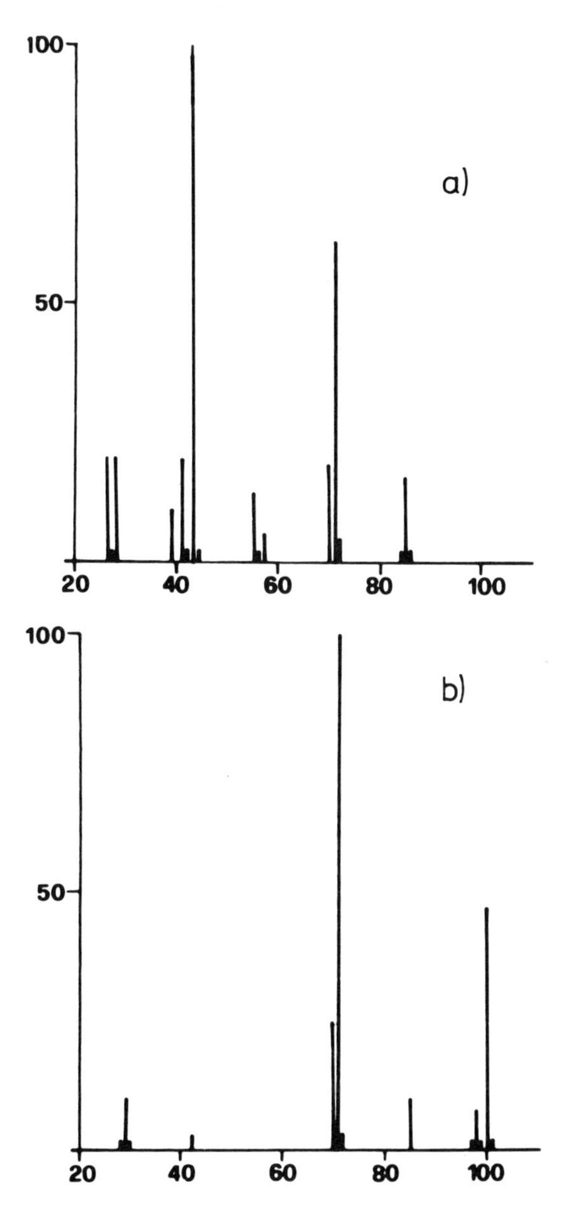

Figure 20 Comparison of (a) the electron bombardment and (b) the field ionization spectra of 3,3-dimethyl-pentane.

The concentration of the substance is about 1%, and the gas pressure is 1−2 mm Hg. Under these conditions practically all the electrons ionize the methane molecule. The primary reaction is to give various ions derived from the methane molecule.

$$CH_4 + e^- \longrightarrow CH_4^+, CH_3^+, CH_2^+, CH^+, C^+, H_2^+, H^+ \quad (2)$$

These reactive species undergo a secondary reaction with another molecule of methane by the ion-molecule reaction to give

$$\begin{aligned} CH_4^+ + CH_4 &\longrightarrow CH_5^+ + CH_3 \\ CH_3^+ + CH_4 &\longrightarrow C_2H_5^+ + H_2 \\ CH_2^+ + CH_4 &\longrightarrow C_2H_4^+ + H_2 \\ &\longrightarrow C_2H_3^+ + H_2 + H \\ CH^+ + CH_4 &\longrightarrow C_2H_2^+ + H_2 + H \end{aligned} \quad (3)$$

These products can give a tertiary ion-molecule reaction with methane.

$$\begin{aligned} C_2H_5^+ + CH_4 &\longrightarrow C_3H_7^+ + H_2 \\ C_2H_3^+ + CH_4 &\longrightarrow C_3H_5^+ + H_2 \\ C_2H_2^+ + CH_4 &\longrightarrow \text{polymer} \end{aligned} \quad (4)$$

The ions CH_5^+, $C_2H_5^+$, and $C_3H_5^+$ are the most abundant species. The ions have the character and behavior of Lewis' acids/bases and can react with the neutral molecules of the sample, extracting or tranferring a proton. Therefore the mass spectrum displays the products of the reaction between the ionic species of methane and neutral molecules of the sample.

The molecular ion $M^{+\cdot}$ is not obtained, but there can be either an $(M+1)^+$ or $(M-1)^+$ ion, depending on the movement of a proton, as shown in (5).

$$CH_5^+ + R{-}H \longrightarrow RH_2^+ + CH_4 \text{ or } CH_5^+ + R{-}H \longrightarrow R^+ + CH_4 + H_2 \quad (5)$$

Figure 21 shows the mass spectrum of fenfluramine after chemical ionization (a), compared with the spectrum obtained by electron impact (b).

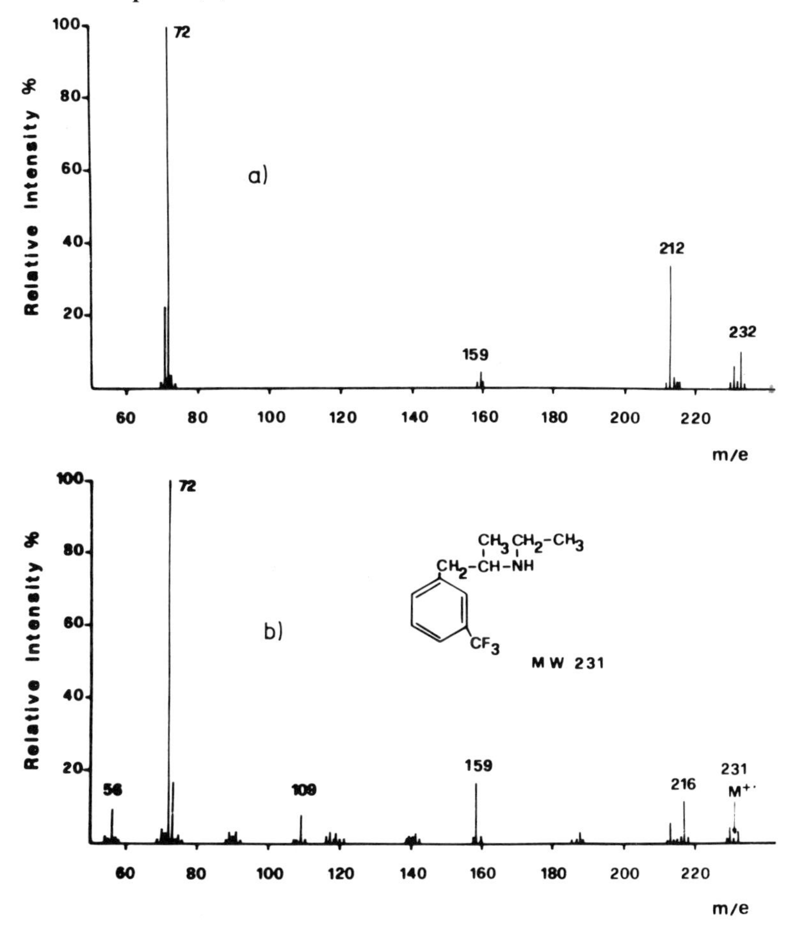

Figure 21 Comparison of (a) the chemical ionization and (b) the electron impact spectra of fenfluramine.

By the chemical ionization method a molecular ion is not formed, but there is an $(M + 1)^+$ ion at $^m/e$232. In chemical ionization spectra, the peaks below $^m/e$ 40 do not have any significance for the compound under examination because they are due to the ions derived from methane.

It is possible in introduce the sample, as the effluent of the gas chromatograph, directly into the ionization chamber if the reagent gas (methane) is also used as the carrier gas.

Simplicity, high sensitivity, and molecular species in the mass spectrum are the main advantages of the system.

This method is limited to volatile substances; it is also limited by the high gas pressures required, which can give rise to electric discharges under the accelerating high tension in the ion source.

B. ISOTOPIC ABUNDANCE METHOD.

All elements (except iodine, fluorine, and phosphorus) possess natural isotopes, and their relative abundances are known. For example, if the mass spectrum of atomic sulphur, S, is carried out, four peaks corresponding to $^{m}/e$ values of 32, 33, 34, and 36 are obtained, and every peak has an intensity proportional to the percentage of the isotope.

For an organic molecule, the molecular peak is always accompanied by the satellite peaks M+1, M+2, M+3, etc. The peak M+1 is due to those molecules that contain an isotope heavier by one mass unit with respect to the most abundant isotope. The M+2 peak is due to those molecules that contain two isotopes heavier by one mass unit or one isotope heavier by two mass units, etc. The intensity of the M+1, M+2 peaks etc. depend on the species, on the number of isotopes present in the molecule, and on their natural abundance.

Consider, for example, the methane molecule, CH_4. Because the abundance of ^{13}C is 1.1% of ^{12}C, the molecular ion of methane at $^{m}/e$ 16 is accompanied by a peak at $^{m}/e$ 17 that is 1.1% of the former (the deuterated ion is neglected because deuterium is relatively much less abundant).

For ethane, CH_3-CH_3, the molecular peak at $^{m}/e$ 30 is accompanied by a peak at $^{m}/e$ 31 that is about 2.2% as intense as the former because there are two atoms of carbon in the molecule.

In general, for a substance containing, C, H, N, and O, a formula (6) exists to calculate the percentage abundance of the M+1 ion relative to M.

$$\frac{(M+1)}{M} = \frac{Wc}{(100-c)} + \frac{Xh}{(100-h)} + \frac{Yn}{(100-n)} + \frac{Zo_1}{(100-o_1-o_2)} \quad (6)$$

where W, X, Y, and Z are the number of atoms of C, H, N, and O respectively and c, h, n, o_1, and o_2 are the natural abundances of ^{13}C, ^{2}H, ^{15}N, ^{17}O, and ^{18}O respectively.

Similarly, a formula exists to calculate the ratio $^{M+2}/M$, but it is long and complex. There are tables that contain all the combinations of the elements (that is, empirical formulae), and the relative abundances of M+1 and M+2 determined by the methods given above, for any given molecular weight. By comparing the intensities reported in the tables with those measured on the spectrum an empirical formula can be deduced. Suppose an unknown substance has a molecular weight of 136 and the intensities measured for the M+1 and M+2 peaks are 11.1% and 0.57% respectively; from the tables it can be determined that the possible empirical formulae are as reported in Table 6.

Table 6

Compound	(M+1)%	(M+2)%
$C_{11}H_{14}$	11.95	0.65
$C_{10}H_{16}$	11.06	0.55
$C_{10}O$	10.84	0.73
$C_9H_{12}O$	9.96	0.64

The formulae $C_{10}H_2N$ and $C_9H_{14}N$ are possible, but need not be considered for the following reasons:

1. If a substance contains only carbon and hydrogen atoms it will have an even numbered molecular weight.

2. If it contains carbon, hydrogen, and oxygen or sulphur (even mass and even oxidation state), the molecular weight is even.

3. If it contains carbon, hydrogen, and phosphorus (odd mass number and odd oxidation state), the molecular weight is even.

4. If, in every case 1., 2., and 3., it contains nitrogen, either (a) the molecular weight will be even if the number of nitrogen atoms is even, or (b) the molecular weight will be odd if the number of nitrogen atoms is odd.

Therefore from the data obtained and by obvious chemical considerations the substance of molecular weight 136 has the empirical formula $C_{10}H_{16}$.

Difficulties arise because of the errors encountered in measuring the intensities of the mass spectral peaks, especially if they are small. In addition, the method is less effective for mass values above $^{m}/e$ 250.

In an approximate but rapid manner the ratios, $^{M+1}/M$ and $^{M+2}/M$, can be determined in the following way.

$$^{M+1}/M = \left(1.1\% \times \text{no. of carbon atoms}\right) + \left(0.36\% \times \text{no. of nitrogen atoms}\right) \quad (7)$$

$$^{M+2}/M = \frac{(1.1\% \times \text{no. of carbon atoms})^2}{200} + \left(0.2\% \times \text{no. of nitrogen atoms}\right) \quad (8)$$

C. ACCURATE MASS MEASUREMENT BY HIGH RESOLUTION MASS SPECTROMETRY.

In modern high resolution instruments the mass of an ion can be measured to the fourth decimal place. Therefore, it is possible to distinguish between, for example, the ion $^{m}/e$ 246.1620 and the ion $^{m}/e$ 246.1984 (a difference of 0.0364 mass units).

A high resolution mass spectrometer can establish not only the exact molecular weight to the fourth decimal place but also the empirical formula.

The measurement of the exact mass of an ion of unknown composition can be performed by comparison with an ion of known composition (Peak Matcher) and calculation of the acceleration potentials required to carry the ions to the collector through a constant magnetic field. The "known" ion can be part of the mass spectrum of the same substance, although a suitable compound, called a "reference," is usually introduced.

A supplementary coil, connected to an oscilloscope, is applied to the magnet of the double focusing mass spectrometer. On the oscilloscope a portion of the mass spectrum is shown. By varying the voltage of the coil applied to the magnet, the instrument can be focused on a single peak—for example, the ion of known composition. While the magnetic field is held constant, the acceleration potentials is varied until the peak of unknown composition superimposes exactly on that of the known ion. If mass m_1 of the ion of known composition and the acceleration potentials v_1 and v_2 of the known and unknown ions respectively are known, the mass m_2 of the ion of unknown composition can be calculated from the simple relationship

$$m_1 v_1 = m_2 v_2 \tag{9}$$

The accuracy of the determination depends on the difference between the mass of the known ion and that of the unknown ion —that is, the accuracy increases as the difference decreases, and in general the difference must not be greater than 10%.

For example, consider a mixture of four compounds having a molecular weight of 260. It consists of

tridecylbenzene $C_{19}H_{32}$
1,2-dimethyl-4-benzoylnaphthalene $C_{19}H_{16}O$
Phenylundecylketone $C_{18}H_{28}O$
2,2′-naphthylbenzothiophene $C_{18}H_{12}S$

The mass spectrum of the peaks corresponding to the molecular ions is given in figure 22.

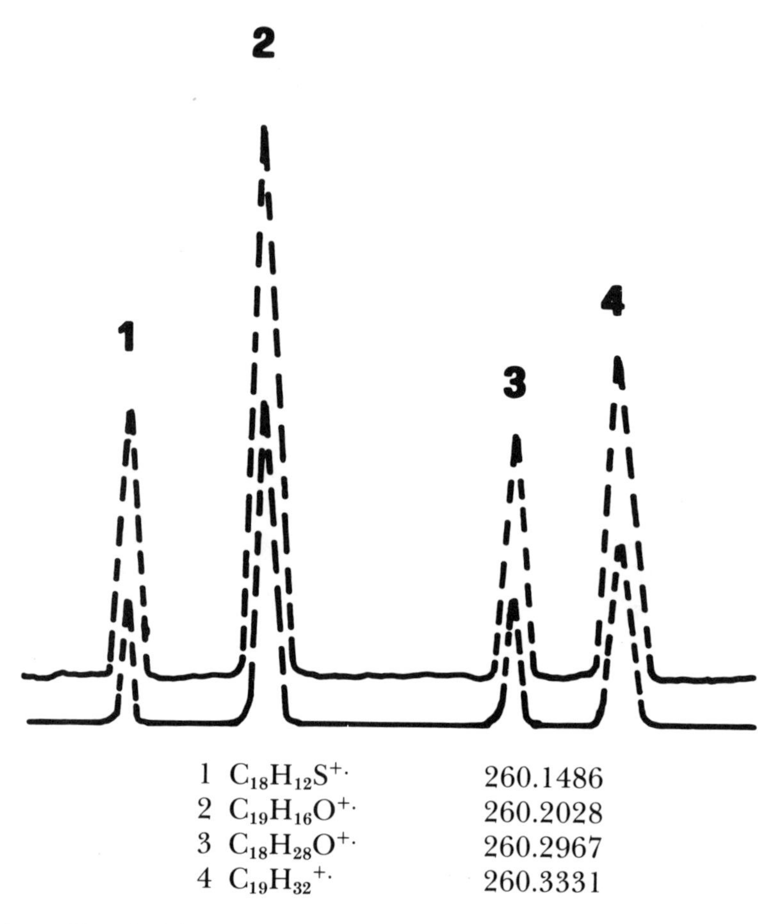

Figure 22 The high resolution mass spectrum of four ions of nominal mass 260.

The determination of the masses has been performed by comparison with the peak 238 of dibromobenzene, introduced as a reference, and the results are given in the following table 7.

Table 7

ion	measured mass	calculated mass	difference
$C_{18}H_{12}S^{+\cdot}$	260.1482	260.1486	0.0004
$C_{19}H_{16}O^{+\cdot}$	260.2026	260.2028	0.0002
$C_{18}H_{28}O^{+\cdot}$	260.2964	260.2967	0.0003
$C_{19}H_{32}{}^{+\cdot}$	260.3329	260.3331	0.0002

Although the reference peak differs in mass by more than 10% with respect to the "unknown," the third decimal place of the molecular weight can be established with certainty and the maximum error at the fourth decimal place is only ±4.

The determination of the accurate mass of an ion can be performed not only for the molecular ion but also for any ion in the mass spectrum. Thus the elemental composition of the ion can be determined, which facilitates the interpretation of the spectrum and the determination of the structure of an unknown molecule.

CHAPTER IV

A. FRAGMENTATION OF POSITIVE IONS.

At the time of the first studies on mass spectrometry it was thought that the organic molecule underwent accidental decompositions following electron bombardment. Such an approach was justified by the fact that the energy of electron bombardment (in the order of hundreds of kilocalories) was very much higher than the energy of activation required in a normal chemical reaction. After a systematic study of an enormous number of compounds by high resolution mass spectrometry, aided by stable isotopic substitution, by metastable transitions, and by the accurate mass measurements of an ion with this system, it has been observed that the decompositions follow certain general rules, comparable with those that can be used to explain the course of a chemical reaction. The application of mass spectrometry to the elucidation of organic structures is based on the existence and validity of these rules.

Before the beginning the discussion on the fragmentation of positive ions it is useful, first, to see some relative considerations of different types of ions that are formed.

B. IONS WITH ODD OR EVEN NUMBERS OF ELECTRONS.

Organic molecules, without exception, are species with an even number of electrons. Because ionization occurs by the removal of an electron, the molecular ion can be identified with a radical-ion that possesses an odd number of electrons, or rather, an unpaired electron. A molecular ion, or any ion, with an odd number of electrons (i.o.e.) can decompose by homolytic of heterolytic processes. The homolytic cleavage occurs by the transfer of a single electron, and this is indicated by the use of an arrow with one hook ⌒

$$CH_3{-}CH_2{-}\overset{+}{\dot{C}}H{-}\dot{C}H_2 \longrightarrow \dot{C}H_3 + CH_2{=}CH{-}\overset{+}{C}H_2 \leftrightarrow \overset{+}{C}H_2{-}CH{=}CH_2 \quad (1)$$

The heterolytic cleavage occurs by the transfer of a pair of electrons and is indicated by a conventional arrow ⌒

$$CH_3{-}CH_2{-}\overset{+\cdot}{Br}: \longrightarrow CH_3{-}\overset{+}{C}H_2 + :\dot{Br}: \quad (2)$$

In both cases, an ion with an even number of electrons (i.e.e.) is obtained by the elimination of an uncharged fragment with an odd number of electrons (radical, r).

When there are two consecutive or simultaneous fragmentations of an ion with an odd number of electrons, an ion with an odd number of electrons (i.o.e.) and a neutral molecule with an even number of electrons (n.e.e.) are formed by either type of cleavage (homolytic or heterolytic).

$$R{-}CH(H){-}CH_2(\overset{\cdot +}{O}H) \longrightarrow R{-}\dot{C}H{-}\overset{+}{C}H_2 + H_2O \quad (3)$$

The majority of ions in a spectrum have an even number of electrons. Either they decompose by loss of a non-charged fragment or a neutral molecule with an even number of electrons to give an ion with an even number of electrons, or they decompose by loss of a radical to give an ion with an odd number of electrons. The latter fragmentation is very rare.

$$CH_3-CH_2-\overset{+}{C}H_2 \longrightarrow \overset{+}{C}H_3 + CH_2{=}CH_2 \qquad (4)$$

$$CH_3-CH_2-\overset{+}{C}H_2 \longrightarrow \dot{C}H_3 + \dot{C}H_2-\overset{+}{C}H_2 \qquad (5)$$

The ions with an even number of electrons give rise to a significant series of peaks for many types of compounds, particularly in the lower part of the spectrum. The aliphatic hydrocarbons form two series of alkyl ions, $C_nH_{2n}^{+\cdot}$, $C_nH_{2n+1}^{+}$, the aliphatic amines a series of $C_nH_{2n+2}N^+$, the alcohols a series of $C_nH_{2n+1}O^+$, the ketones a series of $C_nH_{2n}O^{+\cdot}$. In all cases the series are useful for recognizing the class to which a substance belongs.

The concepts expressed above are summarized in table 8.

Table 8

i.o.e.	⟶	i.e.e.	+	r.	simple homo- and heterolytic cleavage
i.o.e.	⟶	i.o.e.	+	n.e.e.	double cleavage
i.e.e.	⟶	i.e.e.	+	n.e.e.	simple heterolytic cleavage
i.e.e.	⟶	i.o.e.	+	r.	simple homolytic cleavage

C. FACTORS INFLUENCING THE FRAGMENTATION OF AN ION.

The mass spectrum of a molecule consists of numerous peaks, some of which are intense, whereas others are weak or scarcely visible. The preferential formation of some ions depends on the molecule's tendency to break some bonds rather than others and on the stability of some fragments because of their particular structure. In a discussion of the reasons for the formation of ions it is necessary to consider three main factors.

1. The relative strength of the bonds.
2. The stability of the fragmentation products. The products are of three types: neutral molecules, radicals, and positive ions. Generally it is the latter that are the most important factor in directing the fragmentation.
3. The relative spatial arrangement of the atoms or groups.

The three factors are structurally correlated, and it is difficult to establish which is predominant in the course of the fragmentation of a molecule. Very often the stability of the fragmentation product predominates, yet there can be two or more parallel fragmentations for each of the different factors.

1. Relative Strength of the Bonds.

The energies of the most common bonds of an organic molecule are reported below. Obviously the weakest bonds are the single bonds, and therefore in a molecule containing single and multiple bonds, the former are expected to break preferentially.

The weakest bonds are those of the C-X type, where X = Br, I, O, and S. In fact, for the bromides, iodides, ethers, alcohols, thioethers, and thiols, the ions observed are derived from the cleavage of the carbon-heteroatom bond. As will be shown later, sometimes the strength of a bond constitutes the most important factor that rules the fragmentation of a molecule in the mass spectrometer (table 9).

Table 9

Energy of some bonds present in organic compounds.
(Kcal/mole)

Bond	Single	Double	Triple
C-H	97.8		
C-C	82.6	145.1	199.6
C-N	72.8	147	212.6
C-O	85.5	179	
C-S	65	128	
C-F	116		
C-Cl	81		
C-Br	68		
C-I	51		
O-H	110.6		

2. Stability of an Ion.

The electron extracted from a molecule can come from any bond or from a lone pair. However, for an interpretation of the fragmentation mechanism, it is useful to localize the electron extraction on an atom or group having electrons available. For the same reasons, a positive ion can be represented by a positive charge situated in the vicinity of an atom or group that can give it stability, according to the known rules of physical organic chemistry.

From these considerations the fragmentation patterns follow the standard mechanisms of organic reactions, for which the stabilizing factors are the inductive effect, the participation of neighboring electrons or groups, and the resonance effect.

The analogy with organic reaction mechanisms helps to simplify the study of fragmentation patterns, because the exact electronic and nuclear structure of an ion is unknown.

Organic reactions depend on factors such as the interaction with the solvent and the position of equilibrium, whereas the fragmentations of the ions are unimolecular decompositions of strongly activated ions.

2a. Inductive Effect.

In a nucleophilic substitution reaction of the S_N1 type, the comparison between the relative rates of solvolysis, in water, of a substrate of the type R-Br, indicates that a tertiary bromine reacts 100,000 times faster than a secondary bromine, which in turn is ten times faster than a primary bromine.

In a reaction of this type, there is first the formation of an R^+ ion and then the reaction with the solvent.

$$R{-}Br \xrightarrow{H_2O} R^+ + Br^- \longrightarrow R{-}OH + HBr \qquad (6)$$

The tertiary ion is stabilized by the inductive effect of the alkyl groups attached to the carbon atom that carries the positive charge and by the hyperconjugation effect—that is, the delocalization of the electron of a σ C-H bond to form a π bond with an adjacent carbon atom having an empty orbital.

$$-\overset{H}{\underset{|}{\overset{|}{C}}}-\overset{+}{\underset{|}{C}}- \longrightarrow -\overset{H^+}{\underset{|}{C}}{=}\underset{|}{C}- \qquad (7)$$

Similarly, in the mass spectrometer the formation of a tertiary ion is favored with respect to the secondary or primary ions. Comparing the mass spectra of n-butyl and t-butyl alcohol, the ion $C_4H_9^+$ is formed in both cases but is much more abundant with respect to the molecular ion of t-butanol than n-butanol.

In the straight-chain paraffins, the rupture of the alkyl chain occurs in one of the C-C bonds. In a branched paraffin, the fragmentation occurs at the most substituted carbon atom to give the most stability to the ion thus formed. In iso-pentane the fragmentation occurs at the bonds of the carbon atom at the C-2 position, and the ions have the following order of stability.

$$\left[CH_3-CH_2-\underset{|}{\overset{CH_3}{C}}H-CH_3\right]^{+\cdot} \longrightarrow CH_3-CH_2-\overset{CH_3}{\underset{+}{C}}H > \overset{CH_3}{\underset{+}{C}}H-CH_3 > CH_3-\overset{+}{C}H_2 > \overset{+}{C}H_3 \quad (8)$$

The inductive effect also has an influence on the breaking of a bond. In compound R-X, in which X can be Cl, Br, O, S, or N, the electronic attraction on the part of the heteroatom X lowers the electron density of the R-X bond. Formation of an alkyl ion occurs according to reactions (9) and (10).

$$R-\overset{+\cdot}{\ddot{Cl}}: \longrightarrow R^+ + :\dot{\underset{..}{Cl}}: \quad (9)$$

$$R-\overset{+\cdot}{\underset{..}{O}}-R \longrightarrow R^+ + :\ddot{O}R \quad (10)$$

The order of effective electron attraction is Cl > Br, O, S > I ≫ N, C, H.

The unsaturated functional groups, such as the carbonyl group, have a similar effect.

$$R-\overset{:O^{+\cdot}}{\overset{\|}{C}}-R \longrightarrow R-CO^{\cdot} + R^+ \quad (11)$$

Reactions (9), (10), and (11) are represented as heterolytic fragmentations directed by the ionic center present on the heteroatom.

2b. Neighboring Electron Participation.

A carbonium ion of the type $>C^{+}-X$: is stabilized by the presence of an heteroatom that possesses at least a couple of electrons not used in bonding to give the canonical form $>C=X^{+}$.

The stabilization is such that an ion of this type is formed preferentially every time the structure of the molecule permits it. This occurs in alcohols, ethers, thiols, thioethers, and amines.

$$R{-}CH(\overset{\cdot+}{O}H){-}R' \longrightarrow R^{\bullet} + CH(={}^{+}OH){-}R' \quad ; \quad \longrightarrow R{-}CH(={}^{+}OH) + {}^{\bullet}R' \qquad (12)$$

$$R{-}CH_2{-}\overset{+\cdot}{\ddot{O}}{-}CH_2{-}R' \longrightarrow R{-}CH_2{-}\overset{+}{O}{=}CH_2 + {}^{\bullet}R' \quad ; \quad \longrightarrow R^{\bullet} + CH_2{=}\overset{+}{O}{-}CH_2{-}R' \qquad (13)$$

$$R{-}CH(\overset{\cdot+}{N}H_2){-}R' \longrightarrow R{-}CH(={}^{+}NH_2) + {}^{\bullet}R' \quad ; \quad \longrightarrow R^{\bullet} + CH(={}^{+}NH_2){-}R' \qquad (14)$$

The fragmentation reaction is represented as a homolytic reaction directed by the radical center present on the heteroatom. It is indicated by an α cleavage because the bond is adjacent to that between carbon and the heteroatom.

The capacity of a heteroatom to stabilize an adjacent positive charge is very high for nitrogen and diminishes gradually, passing through sulphur, oxygen, and the halogens, for which it is very low.

In the following examples the stabilizing capacity of various heteroatoms, in the same molecule, is compared. The numerical values indicated below the formulae represent the percentage abundance of the CH_2X ions.

$$\underset{100\%}{\underset{NH_2}{CH_2}} \vdots CH_2 \vdots \underset{9\%}{\underset{OH}{CH_2}} \qquad \underset{100\%}{\underset{SH}{CH_2}} \vdots CH_2 \vdots \underset{60\%}{\underset{OH}{CH_2}} \qquad \underset{100\%}{\underset{OH}{CH_2}} \vdots CH_2 \vdots \underset{12\%}{\underset{Cl}{CH_2}} \qquad (15)$$

The α cleavage is the most characteristic fragmentation of amines, alcohols, and ethers, forming ions of fundamental importance for the determination of the structure of these compounds. A table of the influence of a substituent on the relative stability of the ionic fragment formed from ethylene glycol derivatives, expressed in the ratio of their relative abundance, is given in table 10.

Table 10

Molecule	Ion 1	Ion 2	Ratio ½
$HOCH_2-CH_2OH$	$CH_2=\overset{+}{O}H$	$CH_2=\overset{+}{O}H$	1 : 1
$CH_3-CH(OH)-CH_2OH$	$CH_3-CH=\overset{+}{O}H$	$CH_2=\overset{+}{O}H$	9 : 1
$CH_3O-CH_2-CH_2OH$	$CH_2=\overset{+}{O}CH_3$	$CH_2=\overset{+}{O}H$	7 : 1
$CH_3-CH(OCH_3)-CH(OH)-CH_3$	$CH_3-CH=\overset{+}{O}CH_3$	$CH_3-CH=\overset{+}{O}H$	6 : 1
$CH_3O-CH_2-C(CH_3)(OH)-CH_3$	$CH_2=\overset{+}{O}CH_3$	$(CH_3)_2C=\overset{+}{O}H$	1 : 5

The carbonyl compounds, such as ketones, aldehydes, esters, amides, and the corresponding sulphur derivatives, give rise to ions of the $>C^+=\ddot{X}-$ type, stabilized in the canonical form $-C\equiv X^+$.

The fragmentation reaction can be represented by the

homolytic cleavage directed by the radical center present on oxygen.

$$CH_3-\overset{O^{+\cdot}}{\overset{\|}{C}}-CH_3 \longrightarrow CH_3-\overset{O^{+}}{\overset{\|\|\|}{C}} + {}^{\cdot}CH_3 \qquad (16)$$

The ions of the $>C=X^+-$ type are not very common in normal organic chemistry reactions (an example is given by the solvolysis of $CH_3-CH_2-O-CH_2Cl$, which is 1,000 million times faster than that of $CH_3-CH_2-CH_2-CH_2Cl$), whereas acyl ions are considered to be the reactive species in the Friedel-Crafts acylation and in the esterification of highly hindered acids, such as mesetoic acid, in sulphuric acid.

The non-bonding electrons of a heteroatom can also stabilize a positive charge on carbon, as in the case of $C_4H_8Br^+$ and $C_4H_8Cl^+$, which are formed from the halide according to reaction (17).

(17)

2c. Resonance effect.

Vinyl chloride and allyl chloride behave differently toward solvolysis. The first is extremely resistant to nucleophilic reagents, whereas the second reacts by the S_N1 mechanism to form an ionic intermediate stabilized by resonance.

(18)

In the mass spectrometer, the vinylic bonds in the compound of the type $>C=C-C-C-$ or $-C\equiv C-C-C-$ break in a totally different manner from that of allylic bonds.

$$R-CH_2-CH=CH_2]^{+\cdot} \longrightarrow R^{\cdot} + CH_2\text{---}\overset{+}{C}H\text{---}CH_2 \qquad (19)$$

This homolytic type of fragmentation is called β cleavage because it involves the β bond with respect to the functional group.

In mircene (20) the cleavage of the central allylic bond gives the $C_5H_9^+$ ion at $^m/e$ 69, stabilized by resonance.

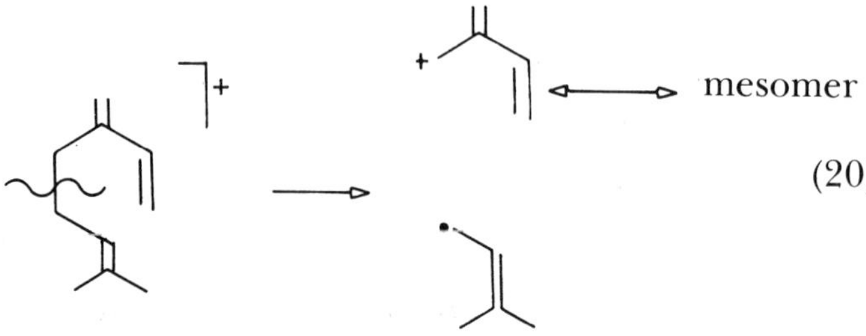

mesomer

(20)

However, in allo-ocumene (21) the ion at $^m/e$ 69 is not formed, because there is no allylic bond in the molecule.

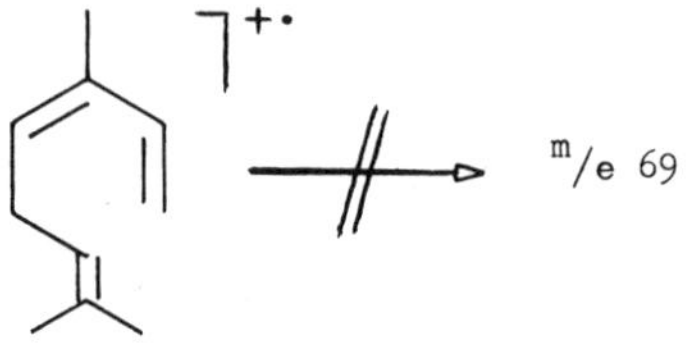

$^m/e$ 69 (21)

The β cleavage is one of the most characteristic fragmentations of aromatic hydrocarbons with an alkyl side-chain. The ion that is formed must be represented by a tropylium structure because all the carbon atoms are equivalent.

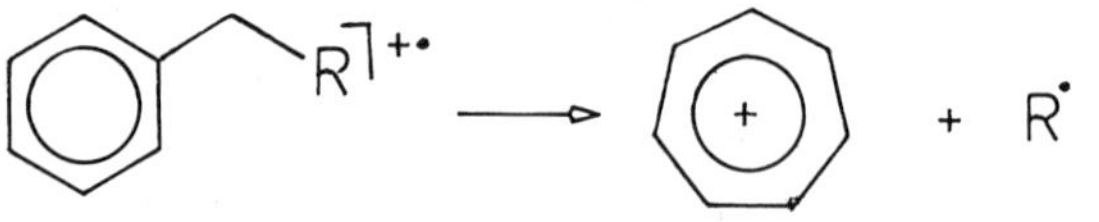

(22)

The β cleavage is also characteristic for five-membered aromatic heterocycles with an alkyl chain in the α or β position. The molecular ion is represented as a positive center with the radical on the heteroatom. The homolytic cleavage is then directed by the radical center to give a common stable ionic species.

The effect of resonance can also influence a fragmentation without acting directly on the stability of the ion.

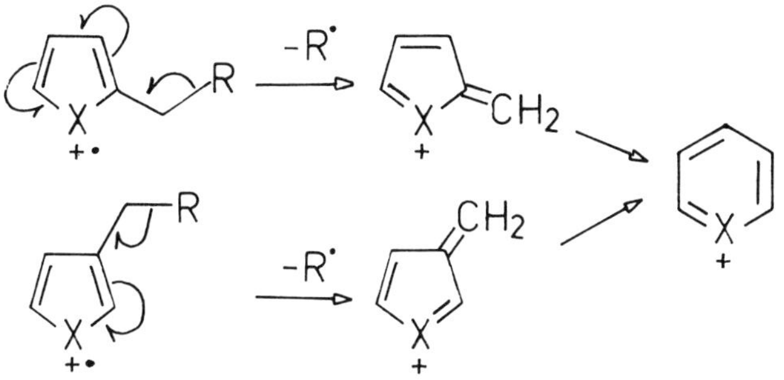

(23)

In the case of acetophenone, benzophenone, benzoic acid, and substituted benzoates, the abundance of the acyl ion, formed as in reaction (24), decreases if the substituent X is an electron-donor, whereas it increases if X is an electron-acceptor.

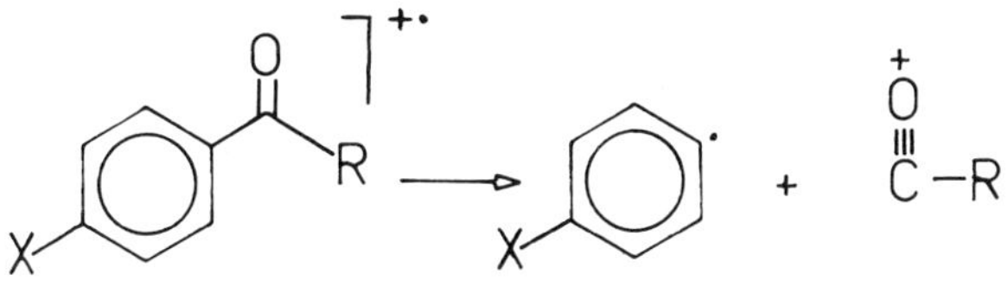

(24)

3. Multi-center Fragmentations and Steric Factors.

The fragmentations so far observed have been simple fragmentations because they involved the cleavage of only one bond. In a complex molecule the interaction of the various functional groups can give a complicated fragmentation reaction that involves the rupture of more than one bond. This is called a multi-center fragmentation.

A multi-center fragmentation is governed by a number of factors: The center involved in the fragmentation must have a suitable arrangement in space; the formation of a stable neutral molecule is the driving force of the process; and the fragmentation occurs by an energetically favorable cyclic transition state.

In common multi-center fragmentation reactions there is a

hydrogen atom migration with the expulsion of a neutral fragment (elimination reactions and McLafferty rearrangement) and the reaction of internal rearrangement of bonds (retro Diels-Alder).

3a. Elimination Reactions.

The elimination process can be represented by scheme (25),

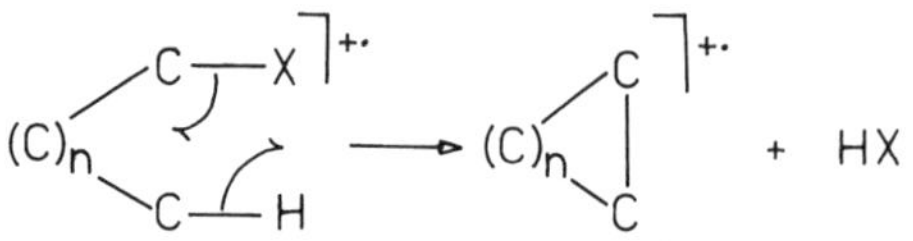

(25)

where X = Cl, Br, I, OH, OR, OCOR, NH_2, NR_2, SH.

The number of carbon atoms between the two groups undergoing elimination (H and X) can vary, but for every atom of X there is a preferred value. In other words, the hydrogen atom that is eliminated must have a precise steric relationship with group X.

The most common elimination reaction is of a water molecule from alcohols to give the M-18 ion. By using a specific substitution of deuterium for hydrogen, it has been demonstrated that in the alicyclic alcohols the reaction does not occur between adjacent groups of atoms, as in the case of chemical elimination, but between carbon atoms in positions 1,3 or 1,4 to form a five- or six-membered cyclic transition state. The reaction products are formed by means of a transfer of a hydrogen atom to the oxygen with homolytic cleavage of the C-H bond, followed by elimination of a neutral molecule (H_2O).

The following scheme (26) shows the case of a 1,4 elimination.

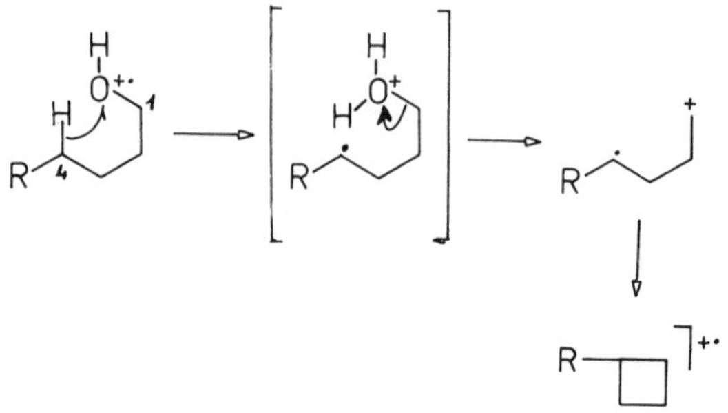

(26)

The M-H_2O ion can be represented by an open chain form, with the positive charge and the radical center on carbons 1 and 4 respectively, or by the cyclic form. Although there may not be valid reasons to support either formula, the cyclic form is preferred because the formation of a new bond in a product lowers the energy required for the reaction.

In a cyclic alcohol the elimination of H_2O has a much more complex route. In cyclohexanol, for example, the introduction of a deuterium atom in positions 2,3 or 4 demonstrates that the eliminated hydrogen atom comes from position 3 or 4, because of the conformation of the ring.

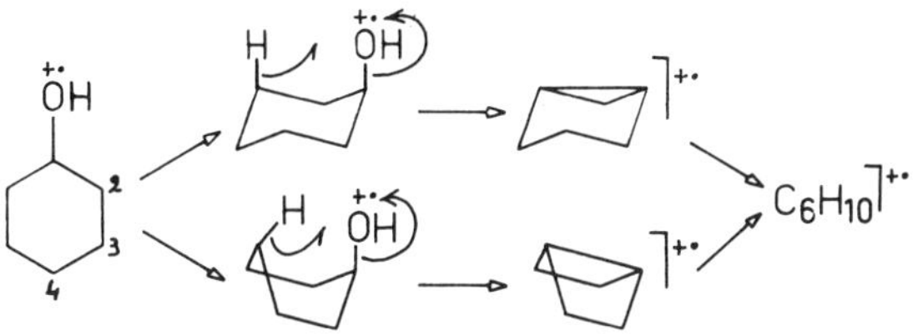

(27)

The tricyclic structures shown do not represent true chemical entities. They are written to assist in the interpretation of the data obtained from the study of the reaction mechanism. It is likely that both, being highly energetic, rearrange to give the common ion $C_6H_{10}^{+\cdot}$

Another very common elimination is that of hydrogen chloride, or hydrogen bromide, which occurs preferentially because of the five-membered cyclic transition state.

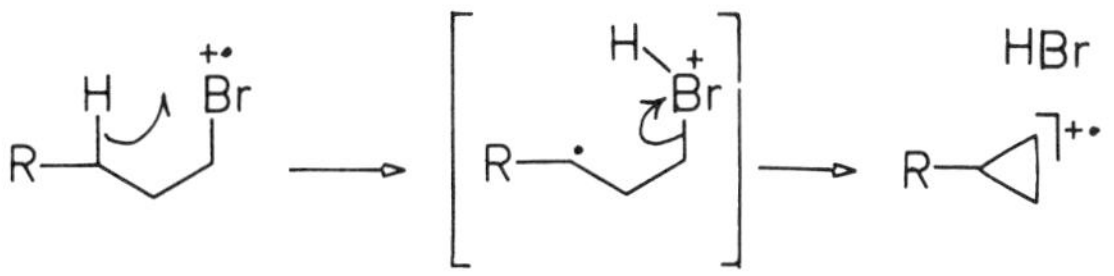

(28)

3b. Ortho Effect.

Ortho disubstituted aromatic systems (29) can give rise to the specific migration of a hydrogen atom onto an atom or group that is eliminated, forming a neutral molecule.

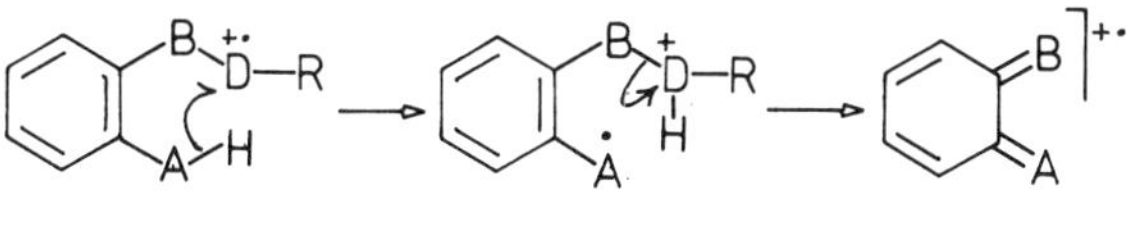

(29)

The atom A can be C, O, N, or S; atom D can be O, N, or S. The neutral molecules eliminated are usually water, alcohols or acids, hydrogen sulphide or mercaptans, ammonia or amines. Besides distinguishing the meta- and para-isomers from those of the ortho disubstituted benzene, this effect differentiates the *cis*- and *trans*-isomers of a suitably substituted olefin. Thus a *cis*-compound of type (30) gives a peak in the mass spectrum that is due to the elimination of methanol, and which will be more intense than that in the spectrum of the *trans*-isomer.

(30)

3c. McLafferty Rearrangement.

The compounds that possess an atom of hydrogen in the γ position with respect to atom A (the site of the positive charge and the radical center) give rise to a specific migration of a hydrogen atom to atom A through a six-membered cyclic transition state, with the elimination of a neutral molecule.

The rearrangement can be shown in two stages, the migration of hydrogen and the expulsion of a neutral molecule, but it really occurs by a single process. The rearrangement represented by reaction (31) is called the primary; the primary can be followed by a second specific migration of hydrogen (32), if the structure of the molecule allows it. This is called a secondary or consecutive rearrangement.

(31)

(32)

The specificity of the hydrogen migration has been demonstrated through the use of deuterated compounds and is attributed to the directional character of the radical center—that is, the non-bonding orbital of atom A, which is occupied by an unpaired electron.

The carbonyl compounds (ketones, aldehydes, esters, amides) are typical examples of substances that can give rise to the McLafferty rearrangement. In this case atom A in the general formula (31) represents an oxygen atom and either the others are all atoms of carbon or one atom is of oxygen or nitrogen and all the others are carbon atoms.

Other molecules that satisfy the structure requirements of the McLafferty rearrangement are the olefins, the alkylbenzenes, the alkylaryl ethers, the nitrogen derivatives of the carbonyl group (oximes, hydrazones, imines), the nitriles, and the epoxides.

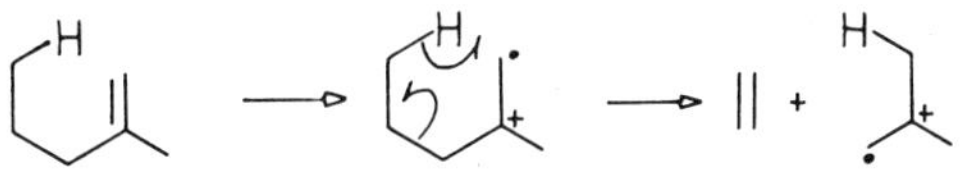

(33)

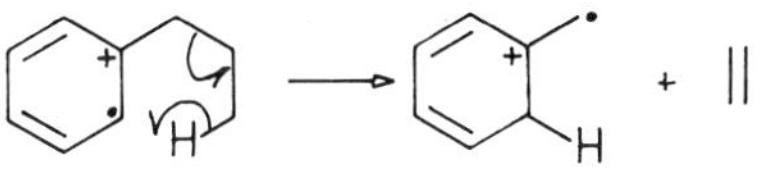

(34)

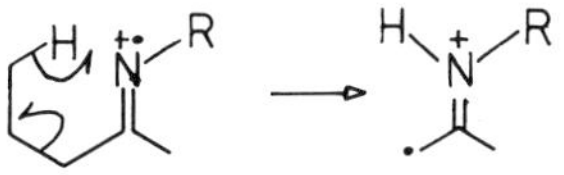

(35)

The McLafferty rearrangement can be influenced by electronic and steric factors. In alkylbenzenes that have an electron donating group in the meta- position, with respect to the alkyl side-chain, the tendency of the specific migration of a hydrogen atom is opposed by the localization of the positive charge on the carbon atom onto which the hydrogen atom must migrate.

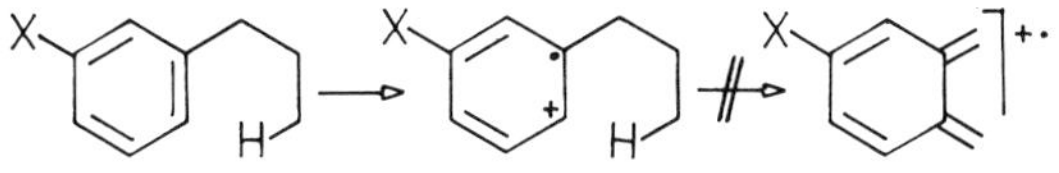

(36)

X = an electron donor

Similarly, in compound (37), in which R = CH_3, the hydrogen migration is sterically hindered by the methyl groups in the ortho-position.

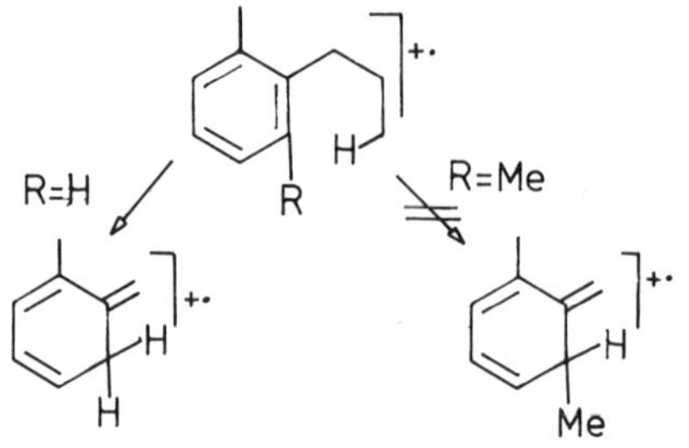

R=H R R=Me H H Me (37)

3d. Retro Diels-Alder Reaction.

The retro Diels-Alder reaction, a multi-center fragmentation, is very common in the mass spectrum of cyclic olefins. Usually, a positively charged diene fragment and a neutral olefinic fragment are formed, but sometimes the olefinic fragment can carry the positive charge and the radical, depending on its structural stabilization factors.

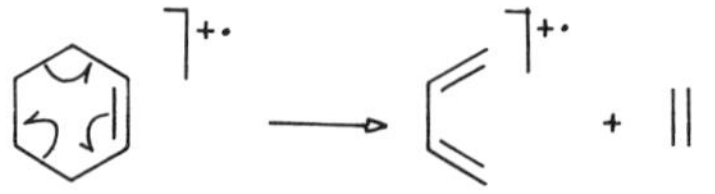

(38)

The mechanism of the reaction can be directed either by the radical center or by the ionic center.

The retro Diels-Alder is of particular importance in the structural determination of numerous compounds, such as norbornene (39), tetraline (40), tetrahydrocannabinol (41), and the triterpenes of the oleananic (42) or ursanic (43) skeleton that have a double bond in position 12,13.

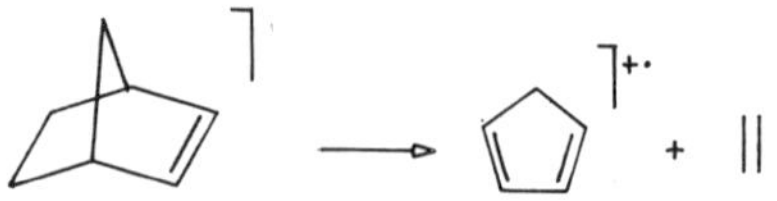

(39)

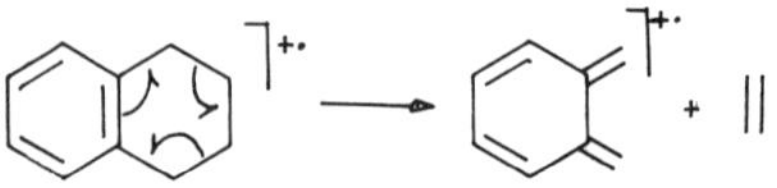

(40)

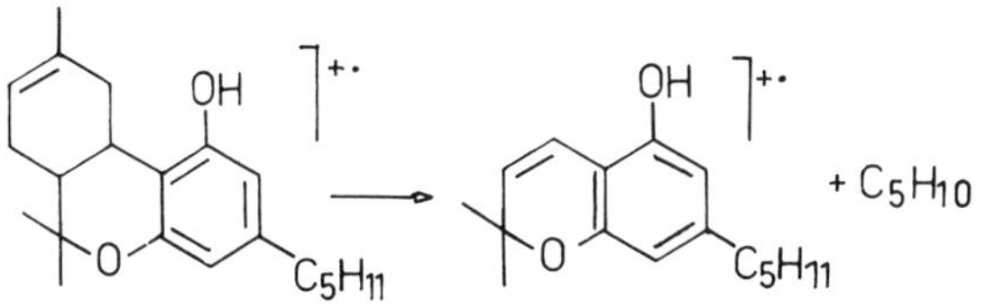

$$+ C_5H_{10} \qquad (41)$$

$$R_1 = CH_3 \qquad R_2 = H \qquad (42)$$

$$R_1 = H \qquad R_2 = CH_3$$

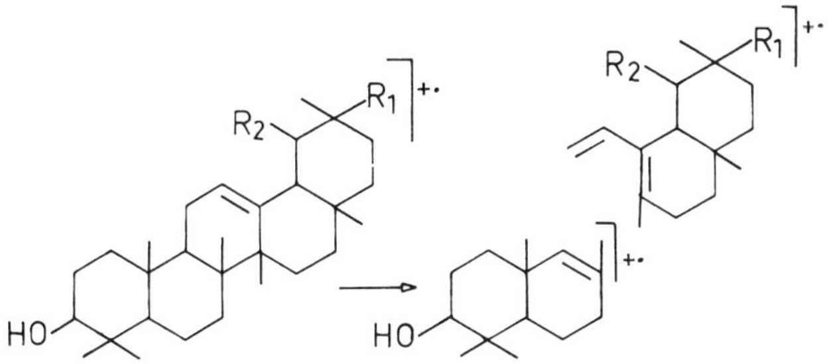

(43)

3e. Hydrogen Migration.

The McLafferty rearrangement and the elimination of HX are examples of the specific migration of hydrogen atoms or groups of atoms with the expulsion of a neutral molecule.

In the mass spectrum of a complex molecule, the migration of a different type from those mentioned above is also a very common fragmentation process. Some processes are not specific, as in the saturated hydrocarbons where there is complete randomization of the hydrogen. The spectrum of neopentane shows, for example, at $^{m}/e$ 29, a peak attributable to an ethyl group that is not present in the molecule.

Other processes have a specific mode of action because they are directed by the radical center and occur in cyclic molecules of type (44), where X = O, OH, OR, NH, NH_2, or NR_2. The cleavage of the molecular ion driven by the radical center on the heteroatom occurs without loss of mass and performs the separation of the positive charge and the radical center. The latter extracts a hydrogen atom through a six-membered cyclic transition state with successive rupture of the bonds adjacent to the new radical center and formation of a stable conjugated ion. Ions of this type give very important peaks in the mass spectra of cyclohexanes, cyclohexanols, and cyclohexylamines.

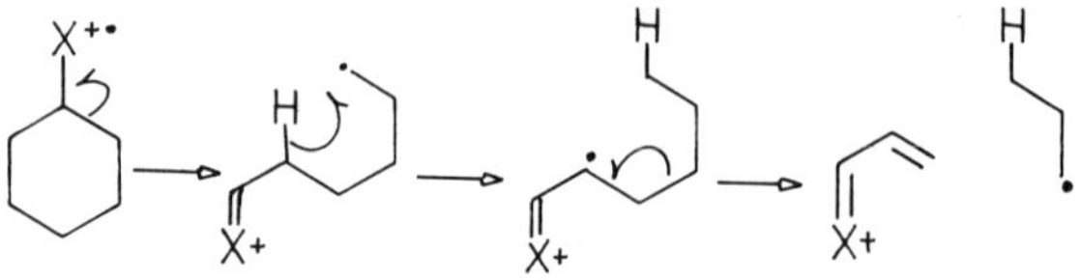

(44)

An analogous fragmentation occurs in the cyclopentane derivatives, in which the transfer of hydrogen takes place through a five-membered cyclic transition state. In the cyclobutane derivatives (45) the hydrogen migration does not occur because it would require the formation of a four-membered cyclic transition state.

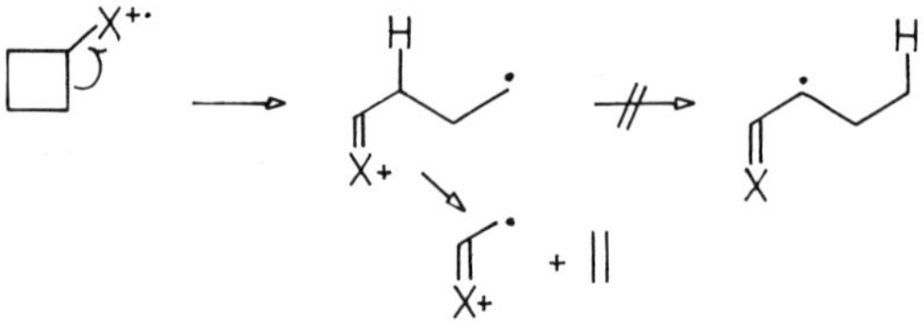

(45)

3f. Expulsion of a Stable Neutral Molecule.

The stability of a neutral fragment has been neglected in the preceding discussion. Nevertheless, it often has a considerable influence on the course of a fragmentation.

The neutral molecules most often eliminated are CO, CO_2, N_2, SO_2, $CH_2{=}CO$, $CH_2{=}CH_2$, and $CH{\equiv}CH$.

The expulsion of ketene and ethylene occurs, in general, through the hydrogen migration reaction, as already shown in the McLafferty rearrangement for ethylene, and as indicated in the following reaction of an acetate to alcohol to form ketene.

$$R{-}\overset{+\bullet}{O}{-}C({=}O){-}CH_2{-}H \longrightarrow R{-}\overset{+\bullet}{O}{-}H \quad O{=}C{=}CH_2 \tag{46}$$

The elimination of the other groups requires the cleavage of two bonds without hydrogen transfer and the formation of a new bond or a lone pair of electrons in the molecule that is eliminated.

Anthraquinone (47) loses two molecules of carbon monoxide to give a compound $C_{12}H_{10}$, to which is attributed the diphenyl structure.

$-CO$; $-CO$ (47)

The mass spectrum of phthalic anhydride (48) shows the elimination first of CO_2, and then of CO to give an ion at $^m/e$ 76, formulated as benzyne. This behavior is analogous to that of the same compound on heating.

$+e$; $-CO_2$; $C_7H_4O]^{+\cdot}$; $-CO$; $C_6H_4]^{+\cdot}$; $^m/e$ 76 ; Δ ; $+ CO + CO_2$ (48)

Acetylene is eliminated from aromatic compounds. Benzene, for example, loses a molecule of acetylene, but it is not possible to formulate a reliable mechanism for this reaction. It seems that the molecular ion of benzene should be represented by the open chain form and not the cyclic form.

$$]^{+\cdot} \longrightarrow C_4H_4]^{+\cdot} + HC{\equiv}CH$$
$$\downarrow$$
$$C_2H_2]^{+\cdot} + HC{\equiv}CH \qquad (49)$$

During the fragmentation of some complex molecules, some ions show the loss of a molecule of hydrogen. Generally the driving force of this process is the formation of a stable aromatic structure.

Ion (51), formed from ibogamine (50) as shown in the scheme, loses two atoms of hydrogen immediately to give ion (52), which is highly stabilized by conjugation.

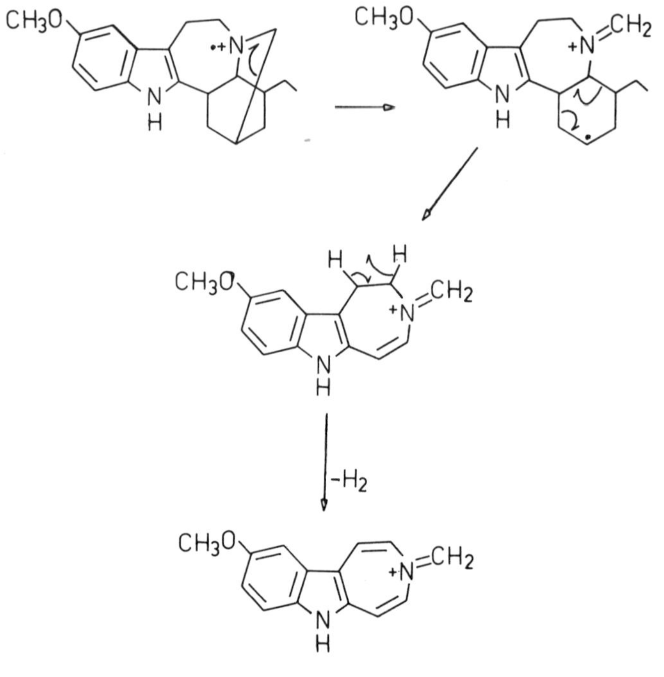

(50)

(51)

(52)

D. INTERPRETATION OF THE MASS SPECTRUM OF AN UNKNOWN STRUCTURE.

The first operation that must be carried out to interpret a mass spectrum is the "counting" of the spectrum.

The mass spectrum is recorded on a strip of photosensitive paper, and it presents the peaks on three traces, of relative intensity 1:10:100. Such traces do not possess an intrinsic zero, and the distance between peaks of successive mass number (the unit scale of the abscissa) is not constant, but rather proportional to $(^m/e)^{1/2}$. Thus the peaks of mass 25, 36, 49, 64, and so forth are equally spaced. Therefore, to determine the value of $^m/e$ of every peak, it is necessary to single out some of the characteristic peaks present in the spectrum, to attribute a mass number to them, and to count from these peaks, peak for peak, up to the molecular peak. The characteristic peaks are those due to carbon(C)$^m/e$ 12, methyl(CH_3)$^m/e$ 15, water(H_2O)$^m/e$ 18, and nitrogen(N_2) and oxygen(O_2), $^m/e$ 28 and 32 respectively.

The identification of these peaks is only apparently difficult, and it is possible, with a little practice, to recognize them immediately without error.

Some instruments possess an accessory, called a "mass marker," that marks intervals corresponding to every $^m/e$ unit directly onto the photosensitive paper, thus giving an immediate count of the spectrum.

Some substances can give rise to peaks at very high mass numbers (above 600), and in these cases it is very easy to make errors in numbering the mass spectral peaks because the peaks are not well-defined. There is a decrease in resolution, and there may be wide "dead" zones deprived of peaks. Therefore a mixture of fluorinated hydrocarbons (called fluolube), that show a series of peaks at 50 unit intervals, is introduced into the mass spectrometer together with the substance. The mass of the peaks of a substance can then be calculated simply by interpolation.

After having "counted" the spectrum, it is useful to mark the points of the most intense peaks so that they are immediately evident whenever the spectrum is inspected, because the trace of a spectrum recorded on photosensitive paper is bound to fade away with time and the inevitable exposure to radiant light.

To conserve a spectrum without fear of it fading, it is better to transfer it onto millimeter graph paper. The most common form of graph has an abscissa of $^m/e$ values and an ordinate of relative intensity—that is, the intensity of the various peaks relative to the base peak, which is the most intense peak in the spectrum and has an arbitrary intensity of 100.

A less common graphic form is that which expresses the intensity of each peak with respect to the sum of the intensity of all the peaks ($\%\Sigma_m$). Thus the intensity of a peak indicates the importance of an ion derived by decomposition of the molecular ion. In the calculation of the total intensity, those peaks for which the intensity is less than 0.5% of the most abundant peaks are neglected. Also, the sum of the intensity usually starts from an ion "m" at $^m/e$ 40, because the peaks at $^m/e$ 32, 28, and 18 are due, mainly, to air and traces of humidity.

The spectra reported by this method are particularly useful for comparing the spectra of isomers in which the only significant differences are quantitative and not qualitative.

On the mass spectrum and its graphic representation the metastable peaks should be indicated with as precise a numerical value as possible, at least to one decimal place.

The metastable ions are fictitious ions the derivation of which can be explained quite quickly. Suppose an ion of mass m_1 is accelerated out of the ion source with a velocity v_1 and that before it enters the magnetic field analyzer it decomposes to give an ion of mass m_2, the latter will be deflected with velocity v_1. This particular ion is not recorded as either m_1 or m_2 but at a mass m*, according to the relationship,

$$m^* = \frac{m_2^2}{m_1} \qquad m_2 < m_1$$

Because it requires about 10^{-6} seconds for the ion to be accelerated and 10^{-5} seconds to enter the analyzer, the ions that have an average lifetime between these two values give rise to metastable peaks. A metastable peak is usually a wide peak of fractional mass and with an intensity of 1−3% of the base peak. The discovery of a metastable peak (m*) is very important in the study of the fragmentation of a substance because it is the only demonstration that the ion m_2 is derived from ion m_1.

After the metastable peaks are singled out and mass numbers are attributed to them, it is important to find the ions m_1 and m_2 that are responsible for the metastable ion. A useful empirical rule for relating m*, m_1, and m_2 is that the distance (in mm) between the metastable ion (m^*) and the ion m_2 is equal to the distance (in mm) between the ions m_1 and m_2.

Generally the molecular ion corresponds to the highest mass. It has been stated previously that not all substances give rise to molecular peaks and that in this case the molecular weight must be determined from the fragment ions, or from the empirical formula established by elemental analysis, or from the presence of certain functional groups obtained by other methods (IR, UV, NMR).

In spectrum no. 1, reported later (fig. 23), the highest peak is mass $^m/e$ 101. Nearby it there is a peak at $^m/e$ 98 that differs by 3 mass units from the former. A fragmentation process giving two separate ions 3 mass units apart is extremely rare (it would require dehydrogenation followed by elimination of a hydrogen radical).

For this reason, and because the substance shows a band at 3,400–3,500 cm^{-1} in the IR (which indicates the presence of a hydroxyl group), a molecular weight of 116 is attributed to the substance. The ions at $^m/e$ 98 and 101 are therefore due to the loss of water and a methyl group respectively.

Sometimes there are intense ions of the type M-1, M-2, M-3, and so forth. The loss of an atom of hydrogen is fairly easy in those substances that possess a cyclic system and a hydrogen atom on the carbon atom adjacent to a heteroatom, such as oxygen or nitrogen.

The loss of two atoms of hydrogen usually occurs in dihydroaromatic systems that undergo a dehydrogenation when the sample is in the inlet system of the spectrometer.

Peaks at intervals of M-5 or M-13 are virtually impossible in normal organic substances, and they are usually due to impurities.

Sometimes there is an M+1 ion instead of the molecular ion $M^{+\cdot}$, derived by a collision between a molecular ion and a neutral molecule. Cases of this type are rather rare but are observed in compounds that contain numerous atoms of oxygen or nitrogen.

Also, the "nitrogen rule" can be a useful aid in establishing whether or not an ion is the molecular ion. As previously explained, the mass of the molecular ion is even if the molecule contains only atoms of C, H, and O or an even number of N atoms. The molecular weight is odd if the molecule contains an odd number of N atoms.

If the molecule contains a chlorine, bromine, or sulphur atom, this is shown by the satellite peaks of the molecular ion and other characteristic ions in the mass spectrum.

Iodine, ^{127}I, and fluorine, ^{19}F, are mono-isotopic, and their presence is revealed by single peaks at $^{m}/e$ 127 and 19 respectively. Chlorine, bromine, and sulphur possess isotopes. The type and the number of these heteroatoms present in a molecule are easily established by the relative intensities of the M, M+2, and M+4 ions.

In spectrum no. 2 (fig. 24), the ions of the doublets at $^{m}/e$ 164,166; 149,151; 107,109; 93,95 are of equal intensity (approximately) and therefore contain a bromine atom.

A comparison between the intensity of the base peak and the molecular peak gives an idea of the overall stability of the ion and the class to which a substance belongs. Thus, for example, the compounds containing extensive aromatic systems, such as the alkaloids of the indole group (yohimbine, heteroyohimbine), harmane, quinoline and furoquinoline, the flavones and isoflavones, and many others, show a molecular peak that is usually the base peak of the spectrum.

In this case it is also very probable that the molecule loses two electrons, after electron impact, and that a doubly charged ion is thereby formed. The aromatic structure has a multi-electron arrangement and can delocalize two positive charges. Pyridine has a system of aromatic electrons and a lone pair and gives rise to a doubly charged ion at $^{m}/e$ 39.5.

At the other extreme are polyhydroxylated substances—for example, the ecdysones. They have a steroidal structure with five or six hydroxyl groups and give rise to a very weak molecular ion. The base peak is an ion that has lost part of the side-chain and has two or three double bonds because of the elimination of water molecules.

To determine the molecular weight and the empirical formula, by elemental analysis, by means of the intensity of peaks M+1 and M+2 or by high resolution, the degree of unsaturation in the molecule must be known—that is, the number of double bonds or rings present in the molecule. For a molecule $C_xH_yO_z$ the degree of unsaturation can be calculated from the formula,

$$x - \frac{y}{2} + 1$$

If there are w atoms of trivalent nitrogen in the molecule, the formula becomes,

$$x - \frac{y}{2} + 1 + \frac{w}{2}$$

If there are halogens present they are counted as atoms of hydrogen.

Now the fragment ions can be considered. In general, the ions derived from an even ion have an odd mass number after simple fragmentation of a bond due to the elimination of a radical. Similarly, the fragments are of even mass if derived from an ion of odd mass number.

If the even ion contains two atoms of nitrogen, the fragment is odd if it contains both N atoms and even if it contains only one. The following example illustrates the point.

$$[CH_3{-}CH_2{-}CH_2{-}CH_2{-}CH{=}N{-}N(CH_3)_2]^{+\cdot}$$
m/e 128

↙ ↘

$$[CH_2{=}CH{-}N{=}N(CH_3)_2]^{+}$$
m/e 85

$$CH_3{-}\overset{+}{N}{-}CH_3$$
m/e 44

If a fragmentation is accompanied by transposition of an atom or group of atoms (elimination of a neutral molecule) from an even mass ion, an even mass ion is formed; those derived from an odd mass number are of odd mass number. If the molecule contains two N atoms, the explanations given above are still true if both N atoms are in the eliminated fragment or both in the charged fragment.

The difference between the mass of an ion and its fragment corresponds to the mass of the fragment eliminated. For a molecule containing C, H, O, and N it is impossible that two ions differ by units between 20 and 25, 31 and 39, or 46 and 57.

The difference must correspond to the mass of a definite chemical entity, either a neutral molecule or a radical.

The derivation of an ion from another must always be explained in terms of the various mechanisms already illustrated, taking into consideration that, following electron impact, profound transformations can occur in the molecule, leading to the elimination of groups of atoms even if these are not present as such in the molecule.

The expulsion of carbon dioxide does not only occur in a molecule containing a carboxyl group. The elimination of water has also been observed in ketones (such as cyclohexanone) and ethers. The loss of a methyl group can be observed in compounds that do not contain such groups (for instance, cyclohexene).

It is a good rule to consider mainly the fragments at high mass numbers and not those fragments at lower mass numbers for the above mentioned possibilities of rearrangements. The molecule neo-pentane does not have an ethyl group, yet in the mass spectrum there is a peak at $^{m}/e$ 29.

The determination of the structure of a substance by mass spectrometry should be confirmed by other spectral data (IR, UV, NMR).

If the substance is known, the spectrum obtained must be qualitatively and quantitatively the same as the reference.

To obtain useful structural information, the mass spectrum of an unknown substance is compared with that of an analogous substance.

It is important to remember that isomers give identical spectra (but different olefins are observed depending on the position of the double bonds, *exo-* and *endo-* isomers of bridged systems). It is rare that the stereo-isomers of simple molecules give different spectra. In complex molecules, however, a stereochemical difference at the center of asymmetry is sufficient to modify the spectrum profoundly.

Some examples of the determination of the structure of organic substances by means of the mass spectrometer are given below.

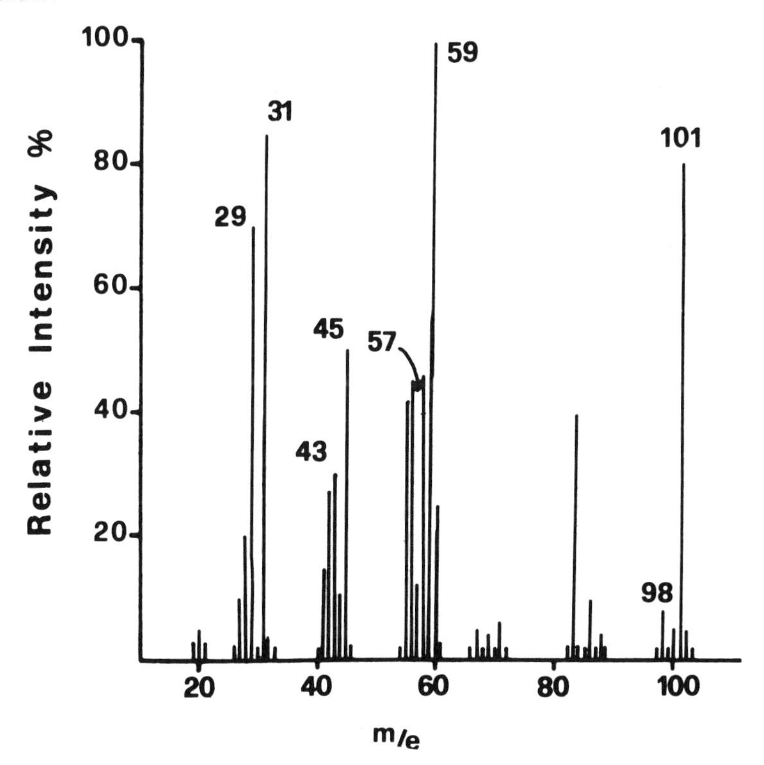

Figure 23 Spectrum no. 1

In the mass spectrum shown in figure 23, the highest peak is at $^{m}/e$ 101, and it would seem to be that of a substance containing an odd number of N atoms. The other peaks of the spectrum belong to two series. One contains only C and H (29, 43, 57), and the other contains C, H, and O (31, 45, 59). None of the important peaks contain nitrogen. This data was obtained by high resolution mass spectrometry. The peak at $^m/e$ 98 cannot be derived from the peak at $^m/e$ 101 and so, as previously stated (p. 69), the molecular peak of this substance is absent. The molecular weight is 116. The ion $^m/e$ 101 is derived, therefore, by the loss of a methyl group and belongs to the series containing C, H, and O, whereas ion $^m/e$ 98 is derived from ion $^m/e$ 116 by the loss of water.

The base peak of the spectrum at $^m/e$ 59 can have one of the two following structures.

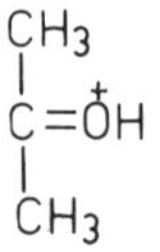

$$(CH_3)_2C=\overset{+}{O}H \quad \text{or} \quad CH_3-CH_2-CH=\overset{+}{O}H$$

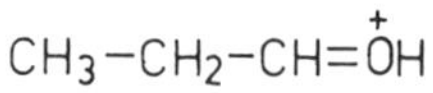

Because there is no peak due to the loss of an ethyl group and because the alcohols give rise to the cleavage of bonds in the β position (with formation of the oxonium ion), the spectrum is that of the alcohol,

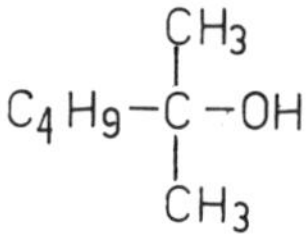

$$C_4H_9-C(CH_3)_2-OH$$

The structure of the C_4H_9- chain is a little difficult to determine. If treated as a branched-chain hydrocarbon of the type

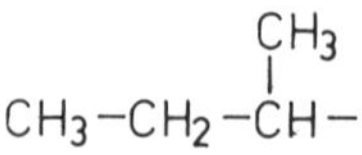

$$CH_3-CH_2-CH(CH_3)-$$

the very intense alkyl ion at $^m/e$ 29 could be derived by the following mechanism.

$$\left[CH_3-CH_2-\underset{|}{\overset{CH_3}{\overset{|}{C}}}H-\underset{CH_3}{\underset{|}{\overset{CH_3}{\overset{|}{C}}}}-OH\right]^{+\cdot} \longrightarrow CH_3-CH_2-\overset{CH_3}{\overset{|}{C}}H-\underset{CH_3}{\underset{|}{\overset{CH_3}{\overset{|}{C^+}}}}$$

$$\longrightarrow CH_3-\overset{+}{C}H_2 + \overset{CH_3}{\overset{|}{C}}H=\underset{CH_3}{\underset{|}{\overset{CH_3}{\overset{|}{C}}}}$$

The alkyl ion at m/e 57 would be due to β rupture with retention of the charge on the alkyl fragment. Because these alkyl ions can also be derived by profound transformations in the molecule — transformations that certainly take place because the ions at m/e 45 ($CH_3CH=\overset{+}{O}H$) and m/e 31 ($CH_2=\overset{+}{O}H$) are formed but are not present in the molecule—it is preferable to leave the structure of the side-chain C_4H_9- as "uncertain" but indicate a probable branched structure, as above.

Confirmation of the structure is obtained by a direct comparison with a sample of 2,2,3-trimethyl-butan-1-ol.

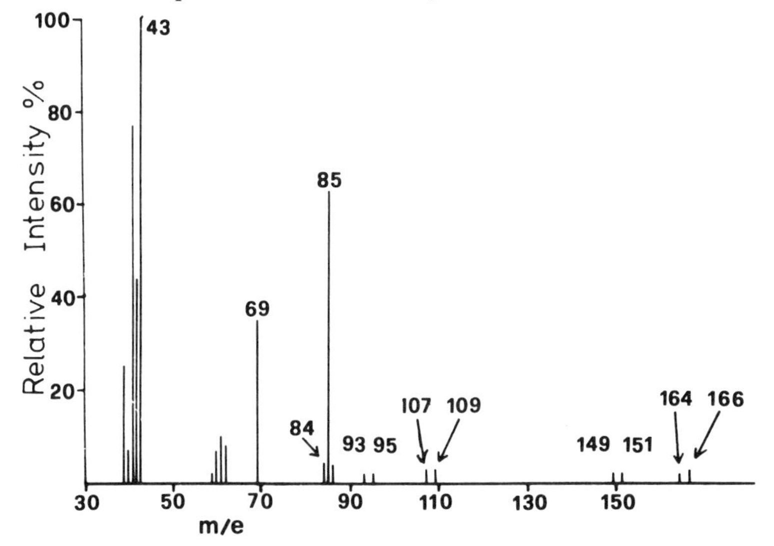

Figure 24 Spectrum no. 2

The intensities of the doublets at $^m/e$ 164,166; 149,151; 107,109; and 93,95 indicate that the molecule contains one atom of bromine (figure 24). Subtraction of the mass of a bromine atom from the mass of the molecular ion gives 164/166 − 79/81 = 85, which corresponds to a fragment containing only C and H $(C_6H_{13})^+$. The peaks at $^m/e$ 93,95 and 107,109 are attributable to the ions $^+CH_2-Br$ and $^+CH_2CH_2-Br$ respectively, showing that the $-CH_2CH_2-Br$ grouping is present in the molecule. Of the four remaining carbon atoms, three must make part of an isopropyl group, which explains the presence of the peak at $^m/e$ 43 as base peak.

The structure of the alkyl bromide is

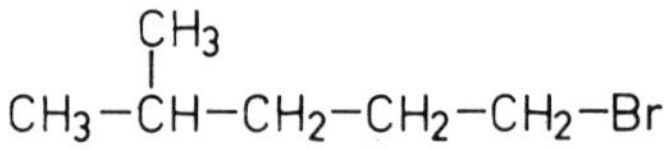

$$\begin{array}{l} \quad\;\, CH_3 \\ \quad\;\;\, | \\ CH_3-CH-CH_2-CH_2-CH_2-Br \end{array}$$

It also explains the formation of the ions at $^m/e$ 149,151 obtained by the expulsion of a methyl group to give the cyclic bromonium ion.

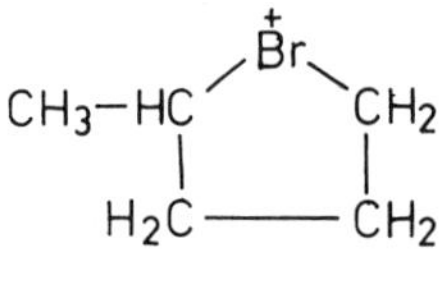

$$\begin{array}{ccccc} & & \overset{+}{Br} & & \\ CH_3-HC & \diagup & & \diagdown & CH_2 \\ | & & & & | \\ H_2C & & \text{———} & & CH_2 \end{array}$$

This behavior is typical for the alkyl chlorides and alkyl bromides. The ion $^m/e$ 84 is due to the loss of HBr and the ion at $^m/e$ 69 to the further loss of a methyl group.

The substance giving spectrum no. 3 (figure 25) contains C, H, and N. The molecular peak is odd, and it is therefore probable that it contains only one atom of nitrogen.

All the fragments indicated are of even mass number and therefore contain the N atom; they are derived by simple fragmentation.

The transition 114→58 is shown by the metastable peak at $^m/e$ 29.5.

The classical fragmentation of the amines is

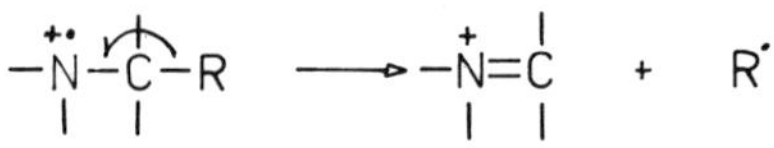

$$-\overset{+\bullet}{\underset{|}{N}}-\underset{|}{C}-R \longrightarrow -\overset{+}{\underset{|}{N}}=\overset{|}{\underset{|}{C}} + R^{\bullet}$$

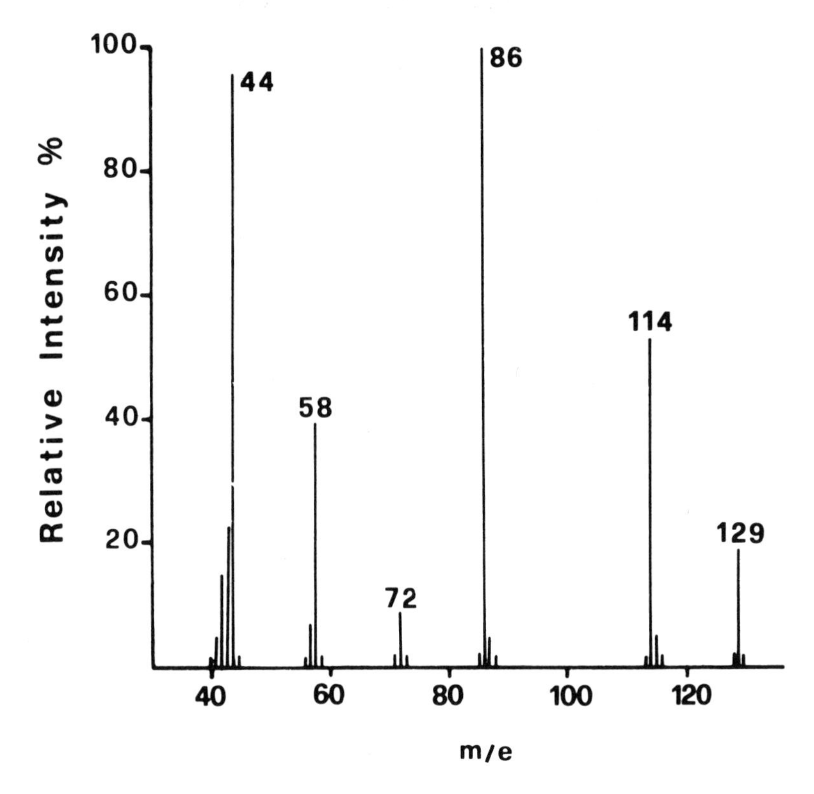

Figure 25 Spectrum no. 3

If this fragmentation gave rise to all the ions, the spectrum will have

for R	$= CH_3$	→	an ion at $^m/e$ 114
	$= C_3H_7$	→	86
	$= C_4H_9$	→	72
	$= C_5H_{11}$	→	58
	$= C_6H_{13}$	→	44

There are, besides an atom of N, eight atoms of carbon, of which one is a methyl group (formation of $^m/e$ 114), three make part of a propyl group (formation of $^m/e$ 86, the base peak), and another is bound to the methyl and the propyl groups. The ions at $^m/e$ 72, 58, and 44 cannot be derived directly from the molecular ion by the expulsion of $C_4H^{\cdot}_9$, $C_5H^{\cdot}_{11}$, and $C_6H^{\cdot}_{13}$ respectively. These ions must be formed by the elimination of a neutral molecule of the type

$$\left[\overset{+\cdot}{\text{-N}}-\text{C}-\text{R} \;\; \cdots \; \text{H}\right] \xrightarrow{-R^{\cdot}} \left[\overset{+}{\text{-N}}=\text{C} \;\; \cdots \; \text{H}\right] \longrightarrow CH_2{=}CH_2 + \overset{+}{\underset{\text{H}}{\text{N}}}{=}\text{C}$$

Because there is no elimination of an ethyl group, the $CH_2{=}\overset{+}{N}H_2$ ion at $^m/e$ 30 is not formed, and the ion $^m/e$ 58 must be derived from ion $^m/e$ 114 (by elimination of C_4H_8), the only possible structure is

$$CH_3-\underset{}{\overset{CH_3}{\overset{|}{C}H}}-\overset{CH_3}{\overset{|}{N}}-CH_2-CH_2-CH_2-CH_3$$

If the metastable ion had not been present at 29.5 it would not have been possible to reach an assignment of the structure, even though this is a very simple case.

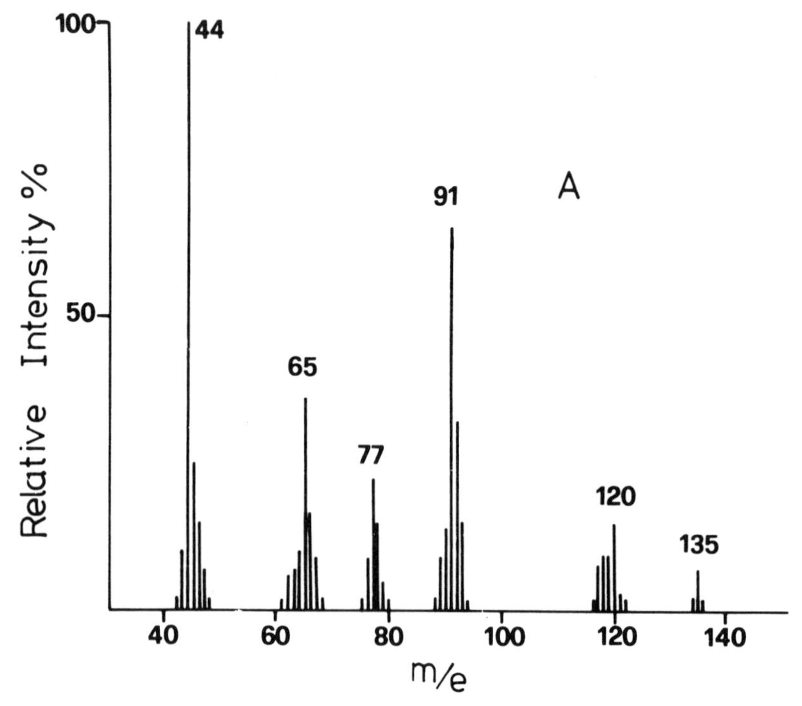

Figure 26 Spectrum no. 4A

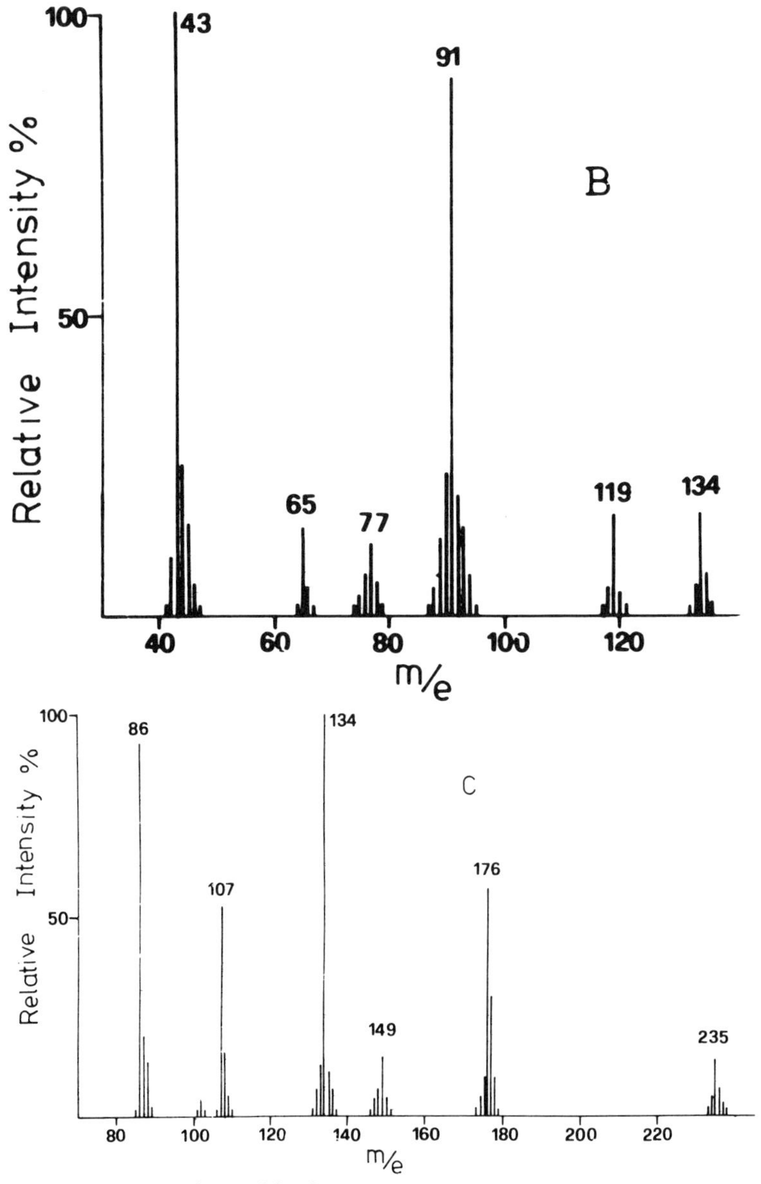

Figure 26 Spectra nos. 4B, and 4C

Spectrum 4A is that of a biological amine, and 4B and 4C are two of its metabolites extracted from urine, after acetylation (figure 26).

The amine A has a molecular ion at $^m/e$ 135 and contains only one atom of nitrogen.

Three peaks in the spectrum are extremely characteristic; those at $^m/e$91, 77, and 65. These peaks are due to a $C_7H_7^+$ ion ($^m/e$91, which loses acetylene to give $C_5H_5^+$, $^m/e$65) and to $C_6H_5^+$ ($^m/e$ 77); they are diagnostic for the presence of the $C_6H_5-CH_2-$ grouping in the molecule.

The facile loss of a methyl group from the molecular ion and a base peak at $^m/e$ 44 indicate that the substance is amphetamine.

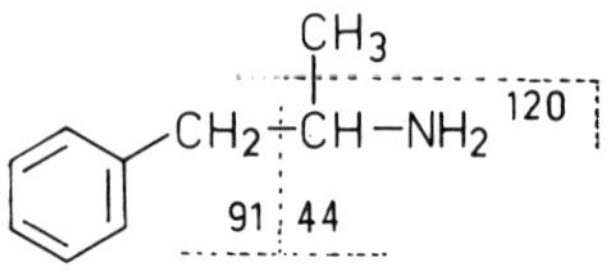

CH3
C6H5–CH2–CH–NH2 120
91 44

The metabolite 4B does not contain nitrogen, but from the spectrum (which is very similar to the former) the structure is deduced to be benzyl methyl ketone.

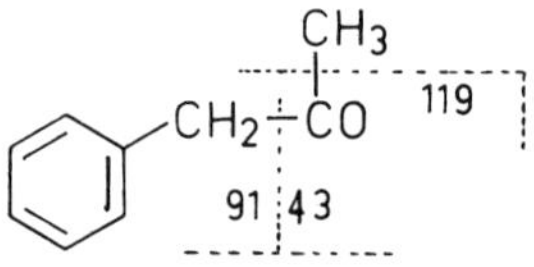

CH3
C6H5–CH2–CO 119
91 43

The metabolite 4C also contains nitrogen but does not show the $C_6H_5CH_2-$ grouping. In its place there is the acetoxy benzyl ion.

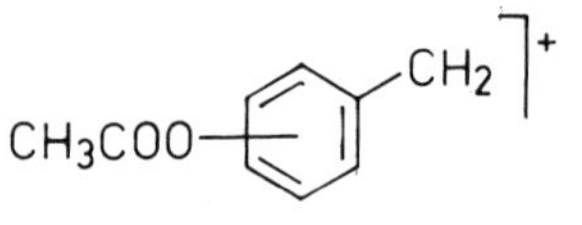

$$CH_3COO-C_6H_4-CH_2^+$$

This arrangement gives peaks at $^m/e$ 149 and 107 (149 – $CH_2{=}CO$, loss of ketene is typical for the acetates of phenols). The remaining part of the molecule gives a peak at $^m/e$ 86 attributable to the ion

$$CH_3-\overset{+}{C}H-NH-CO-CH_3$$

The metabolite 4C must be, therefore, the acetate of a hydroxy-amphetamine.

$$CH_3COO\text{-}C_6H_4\text{-}CH_2\text{-}CH(CH_3)\text{-}NH\text{-}CO\text{-}CH_3$$

The presence of an acetyl group on nitrogen gives rise to a McLafferty rearrangement.

$$\left[CH_3COO\text{-}C_6H_4\text{-}CH(H)\text{-}CH(CH_3)\text{-}NH\text{-}C(=O)\text{-}CH_3\right]^{+\cdot}$$

$$\downarrow$$

$$\left[CH_3COO\text{-}C_6H_4\text{-}CH{=}CH\text{-}CH_3\right]^{+\cdot} \quad + \quad HN{=}C(OH)\text{-}CH_3$$

$^{m}/e$ 176

$$\downarrow -CH_2CO$$

$$\left[HO\text{-}C_6H_4\text{-}CH{=}CH\text{-}CH_3\right]^{+\cdot}$$

$^{m}/e$ 134

In this rearrangement the ionic charge rests on the styrene fragment so as to delocalize the positive charge in the aromatic system.

The transition 176→134 is confirmed by the presence of a metastable peak at $^{m}/e$ 101.8.

It is not possible to establish the position of the acetate group on the benzene ring. Only a comparison with the acetate of p-hydroxy-amphetamine can completely identify the metabolite.

E. COMMON FRAGMENTATION PATTERNS IN DIFFERENT CLASSES OF ORGANIC COMPOUND.

1. Hydrocarbons.

Aliphatic. The molecular peak is always present in the case of compounds with linear chains even if the height decreases with increasing molecular weight. Two series of fragmentation homologues are preponderant in their mass spectra and correspond to the empirical formulae $C_nH_{2n+1}{}^+$ and $C_nH_{2n-1}{}^+$.

The peak corresponding to the loss of a methyl group $(M-15)^+$ is of reduced intensity unless there is a branched methyl group.

The peaks at $^m/e$ 43 = $C_3H_7{}^+$ and 57 = $C_4H_9{}^+$ are abundant.

In the case of branched hydrocarbons the fragmentation takes place at the substituted carbon atoms.

Olefins. The molecular peak $C_nH_{2n}{}^{+\cdot}$ is evident. A series of peaks, at 41 + 14n (n = 0, 1, 2. . .) is observed, corresponding to a series two mass units lower than the most abundant in the aliphatic series. The base peak occurs by allylic cleavage of the molecule.

Saturated Cyclic. Show an intense molecular peak. The side-chain fragments in the α position and the cleavage of the ring give fragments at $^m/e$ 28 = $C_2H_4{}^{+\cdot}$ and 20 = $C_2H_5{}^+$. Because of an abundant series of ions at 27, 41, 55, 69, and 83, it can be difficult to differentiate them from the olefinic hydrocarbons.

Aromatic. The molecular peak is, in general, very abundant, and the peak at mass 77 = $C_6H_5{}^+$ is also evident.

The condensed ring aromatics are extremely stable, and the fragmentation is much reduced and at times nonexistent.

In the alkylaromatics the base peak is that of mass 91 = $C_7H_7^+$, as in the tropylium ion. If, however, the carbon in the α position, with respect to the ring, is substituted, the base peak becomes 91 + 14n.

The positional isomers on the ring (o−, m−, and p−) are difficult to differentiate inasmuch as they give identical transpositions.

2. Hydroxylated Compounds.

Alcohols. The molecular peak is generally weak or nonexistent, particularly in compounds with long aliphatic chains. The peak $(M\text{-}18)^{+\cdot}$ occurs by loss of water from the molecular peak and can be mistaken for the molecular ion.

The molecular ion decomposes by α cleavage with a tendency to lose a radical of large dimensions. The primary alcohols show a base peak at $^m/e$ 31 ($^+CH_2{-}OH \leftrightarrow CH_2{=}\overset{+}{O}H$); the secondary alcohols, with one methyl substituent on the α carbon atom, often have a base peak at $^m/e$ 45 ($^+CHOH{-}CH_3 \leftrightarrow CH_3{-}CH{=}\overset{+}{O}H$); the tertiary alcohols often have a base peak at $^m/e$ 59 ($^+C(CH_3)_2{-}OH \leftrightarrow (CH_3)(_2C{=}\overset{+}{O}H$).

The −OH group does not fragment abundantly. Consequently, in alcohols with a long side-chain, hydrocarbon peaks predominate. The series 31 + 14n (n = 0, 1, 2. . .) is characteristic.

Alcohols with a long side-chain can have $(M-2)^{+\cdot}$ and $(M-3)^+$ peaks, even though they are not very abundant; they are obtained by mechanisms analogous to those of aromatic alcohols, as described below.

Phenols and Aromatic Alcohols. The phenols show an intense molecular peak. The $(M-CO)^{+\cdot}$ and $(M-CHO)^+$ peaks are also intense, whereas $(M-1)^+$ is weak. On the other hand, in the derivatives, such as cresol, the $(M-1)^+$ peak is more abundant than the molecular ion, and a hydroxy-tropylium ion should be formed (cf. toluene).

The aromatic alcohols, by analogy with the aliphatics, show $(M-2)^{+\cdot}$ and $(M-3)^+$ peaks that are explained by the formation of aldehydes and ketones. Moreover, they show the characteristic loss of a single H atom to give hydroxy-tropylium derivatives. For example, benzyl alcohol.

$$C_6H_5{-}CH_2{-}OH^{+\cdot} \xrightarrow{-2H} C_6H_5{-}\overset{H}{\overset{|}{C}}{=}\overset{+\cdot}{O} \xrightarrow{-H^\cdot} C_6H_5{-}C{\equiv}\overset{+}{O}$$

$$\downarrow -H^\cdot$$

$$\text{(OH-tropylium)}^+ \xrightarrow{-CO} (C_6H_7)^+ \xrightarrow{-2H} (C_6H_5)^+$$

3. Ethers.

Aliphatic. The molecular peak is of low intensity. As with the alcohols, there is a characteristic series of peaks at 31, 45, 59, 73, 87, and 101. The charge is carried principally by the oxygen atom after removal of an electron from one of the lone-pairs. There are four major fragments.

1. α cleavage with respect to oxygen ($^+OR = 31 + 14n$).

2. β cleavage with respect to oxygen

$(R{-}\overset{|}{\underset{|}{C}}{-}OR_1 \xrightarrow{-R^\cdot} {>}C{=}\overset{+}{O}R_1 \leftrightarrow {}^+\overset{|}{\underset{|}{C}}{-}OR_1$, $45 + 14n$).

3. α cleavage with respect to oxygen ($R{-}\overset{|}{\underset{|}{C}}{-}OR_1 \rightarrow R{-}\overset{|}{\underset{|}{C}}{}^+$, $29 + 14n$).

4. α cleavage with respect to oxygen, with transposition of a H atom that does not necessarily reside on the carbon atom in the β position relative to oxygen (they migrate from carbon atoms in different positions).

$$\left[R{-}\overset{|}{\underset{|}{C}}{-}O{-}CH_2{-}\underset{H}{\underset{|}{C}H}{-}R'\right]^{+\cdot} \rightarrow R{-}\overset{|}{\underset{|}{C}}{-}OH + \left[CH_2{=}CH{-}R'\right]^{+\cdot}$$

$$(28 + 14n)$$

Other types of cleavage are known in addition to those listed above.

$$R{-}\underset{R_1}{\underset{|}{C}H}{-}\overset{+\cdot}{O}{-}CH_2{-}CH_3 \xrightarrow{-R_1^\cdot} R{-}CH{=}\overset{+}{O}{-}\underset{CH_3}{\underset{|}{C}H_2} \xrightarrow{-C_2H_4} R{-}CH{=}\overset{+}{O}H$$

Acetals. The molecular peak is very weak. The three principal fragmentation patterns are

$$R_1\text{-}C(OR)(OR)\text{-}H^{+\cdot} \xrightarrow{-OR^{\cdot}} R_1\text{-}\overset{+}{C}(OR)\text{-}H$$

$$R_1\text{-}C(OR)(OR)\text{-}H^{+\cdot} \xrightarrow{-R_1} RO\text{-}\overset{+}{C}(H)\text{-}OR$$

$$R_1\text{-}C(OR)(OR)\text{-}H^{+\cdot} \xrightarrow{-H^{\cdot}} RO\text{-}\overset{+}{C}(R_1)\text{-}OR$$

Ketals. There is no molecular ion. They lose an alkyl group, giving a resonance-stabilized oxonioum ion.

$$\xrightarrow{-R_1^{\cdot}}$$

Aromatic. The molecular peak is intense. In general, the base peak, obtained by β cleavage with respect to the ring, follows a transposition.

$$[C_6H_5\text{-}O\text{-}CH_2\text{-}CH_2(H)]^{+\cdot} \longrightarrow C_6H_5\text{-}\overset{+\cdot}{O}H + CH_2{=}CH_2$$

There are also other concurrent reactions,

$$[C_6H_5\text{-}O\text{-}CH_2(H)]^{+\cdot} \xrightarrow{-CH_3^{\cdot}} C_6H_5\text{-}\overset{+}{O} \xrightarrow{-CO} C_5H_5^{+}$$

$$[C_6H_5\text{-}O\text{-}CH_2(H)]^{+\cdot} \xrightarrow{-CH_2=O} C_6H_6^{+\cdot}$$

that in the case of some isomers can become specific.

$$\xrightarrow{-CH_3^{\cdot}} \qquad \xrightarrow{-CO}$$

$$\longrightarrow (M\text{-}OCH_3)^{+}$$

$$\longrightarrow (M\text{-}OCH_2)^{+\cdot}$$

$$\longrightarrow (M\text{-}OCH)^{+}$$

(Double transposition of H to the aromatic nucleus, a rare process.)

4. Epoxides.

Aliphatic. The molecular ion is usually weak. The molecular ion undergoes γ fission, with respect to the heterocyclic ring, giving an intense peak.

$$\left[\text{(epoxide)}-CH_2-CH_2 \vdots CH_2-CH_3\right]^{+\cdot} \longrightarrow \text{(epoxide)}-CH_2-\overset{+}{C}H_2 + \dot{C}H_2-CH_3$$

There can be two types of McLafferty rearrangement, with suitably substituted epoxides.

$$\longrightarrow H\overset{+\cdot}{O}CH_2-CH=CH_2 + CH_2=CHCH_3$$

$$\longrightarrow H\overset{+\cdot}{O}-CH=CH_2 + CH_2=CH-CH_2-CH_3$$

Aromatic. There is an intense M−1 peak. Transannular cleavage gives the tropylium ion, which is intense at $^m/e$ 91.

$$\left[C_6H_5-\underset{H}{C}(-O-)CH-R\right]^{+\cdot} \longrightarrow \text{(tropylium ion, +)}$$

5. Carbonyl Compounds.

Aliphatic Ketones. Show an intense molecular peak. Principal rupture: α ($R \vdots COR_1$ and $R-CO \vdots R_1$), with greater probability of losing the heavier group. The charge is mainly located on the fragment that bears oxygen, with a lower probability of it being on the alkyl chain.

The long alkyl groups can fragment giving a series of 43 + 14n.

An equally important fragmentation is β cleavage followed by a transposition of hydrogen from the γ position (McLafferty rearrangement).

$$\left[R-C(=O)-CH_2-CH_2-CHR_1(H)\right]^{+\cdot} \longrightarrow R-C(=CH_2)-\overset{+\cdot}{O}H + CH_2=CHR_1$$

Aromatic Ketones. There is an intense molecular peak. In general a base peak is obtained by fragmentation α to the carbonyl.

$$C_6H_5-\overset{O^{+\cdot}}{\overset{\|}{C}}-R \xrightarrow{-R^{\cdot}} C_6H_5-C\equiv\overset{+}{O}$$

Aliphatic Aldehydes. The molecular peak $M^{+\cdot}$ and $(M-1)^+$ are intense, the latter following α cleavage. α cleavage ends at propionaldehyde with the formation of the $(H-C\equiv O)^+$ ion at $^m/e$ 29. β cleavage, for linear aldehydes of more than three carbon atoms, with a base peak at $^m/e$ 44 following a McLafferty rearrangement.

$$\left[\overset{1}{H}C(=O)-\overset{2}{H_2C}-CH_2-CH(H)-R\right]^{+\cdot} \longrightarrow HC(=CH_2)-\overset{+\cdot}{O}H + CH_2=CHR$$

If the carbon atom in position 2 is substituted, peaks occur at masses 44 + 14n.

Aromatic Aldehydes. The peaks $M^{+\cdot}$ and $(M-)^+$ are intense. There is a tendency to form the benzoylic cation ($^m/e$ 105) and, in the case of a substituted ring, ions of correspondingly higher mass.

6. Carboxyl Compounds.

Aliphatic Acids. $M^{+\cdot}$ is, in general, irrelevant, but the molecular peak of a long chain monocarboxylic acid is nevertheless visible.

By the McLafferty rearrangement an important peak is formed at m/e ($CH_2{=}\underset{\displaystyle OH}{\underset{|}{C}}{-}OH^{+\cdot}$), which can also be the base peak.

The intense peak at mass 45 is due to (^+COOH). Compounds of lower mass also show characteristic peaks at M−17 (−OH), M−18 (H_2O), and M−45 (−COOH).

To increase the volatility, the methyl esters are usually prepared.

Aromatic Acids. $M^{+\cdot}$ is intense, and so are $(M-17)^+$ and $(M-18)^{+\cdot}$. If the o-position is free there is also an intense $(M-45)^+$ ion.

Esters of Aliphatic Acids. $M^{+\cdot}$ is generally irrelevant. Following α cleavage, with respect to the carboxyl group, two series of fragments are formed depending on the localization of the charge.

$[R{-}CO \vdots OR_1]^{+\cdot} \rightarrow R{-}C{\equiv}O^+ + {\cdot}OR_1$, giving a series of 29 + 14n.

$[R \vdots CO{-}OR_1]^{+\cdot} \rightarrow R^{\cdot} + {}^+O{\equiv}C{-}OR_1$, giving a series of 59 + 14n.

With β cleavage, following a McLafferty rearrangement, the methyl esters give a base peak at m/e 74 ($CH_2{=}\underset{\displaystyle +.OH}{\underset{|}{C}}{-}OMe$), ethyl esters a peak at m/e 88. If R_1, bound to oxygen, is made up of three or more carbon atoms, there is an olefinic transposition that gives peaks (at times intense) at mass R_1-1 (42 + 14n).

The esters of diacids have an intense $M^{+\cdot}$ peak and show a preference for the following fragmentation pathway.

$$\left[RO{-}\underset{\displaystyle O}{\underset{\|}{C}}{-}(CH_2)_n{-}\underset{\displaystyle O}{\underset{\|}{C}}{-}OR_1\right]^{+\cdot} \xrightarrow{-OR_1^{\cdot}} RO{-}\underset{\displaystyle O}{\underset{\|}{C}}{-}(CH_2)_n{-}C{\equiv}\overset{+}{O}$$

$$\left[RO{-}\underset{\displaystyle O}{\underset{\|}{C}}{-}(CH_2)_n{-}\underset{\displaystyle O}{\underset{\|}{C}}{-}OR_1\right]^{+\cdot} \xrightarrow{-COOR_1^{\cdot}} \left[RO{-}\underset{\displaystyle O}{\underset{\|}{C}}{-}(CH_2)_n\right]^{+}$$

Lactones. $M^{+\cdot}$ is very weak. There is the loss of an alkyl radical by α cleavage with respect to the ring.

$$\text{R}\ \text{(lactone ring, } \overset{\cdot+}{\text{O}}\text{, =O)} \xrightarrow{-R^{\cdot}} \text{(ring, } \overset{+}{\text{O}}\text{, =O)}$$

Esters of Aromatic Acids. $M^{+\cdot}$ is intense. In the case of the methyl ester, M−31 (i.e. $M-OCH_3$) is the base peak, and m−59 ($M-COOCH_3$) is also intense.

7. Nitrogenous Compounds.

Aliphatic Amines. $M^{+\cdot}$ is always weak. The base peak is derived by the rupture of a bond between the carbon atoms in the α and β positions, with respect to the N atom. For primary amines, unsubstituted in the α position, the base peak is due to $CH_2{=}\overset{+}{N}H_2$. For secondary amines the base peak results from the loss of a fragment of large mass. The fragmentation takes place at the same time as a transposition of hydrogen in the fragments.

$$CH_2{=}\overset{+}{N}H{-}\underset{\displaystyle CH_3}{\underset{|}{C}H}{-}CH_2{-}CH(CH_3)_2 \longrightarrow CH_2{=}\overset{+}{N}H_2$$

$$CH_3{-}CH_2{-}\overset{+}{N}H{=}CH{-}CH_3 \longrightarrow \overset{+}{N}H_2{=}CH{-}CH_3$$

Therefore all the amines show peaks at mass 30 + 14n.

Aromatic Amines. $M^{+\cdot}$ is intense. In primary amines which do not have bonds that break easily, the ring fragments with the elimination of HCN (M−27) and $H_2CN^{\cdot}$ (M−28) and, in the case of aniline, the probable formation of the cyclopentadiene ion.

The alkyl side-chain breaks, as in the aliphatic amines, at the bond between the α and β carbon atoms, with respect to the N atom.

Aliphatic Amides. $M^{+\cdot}$ is irrelevant. In low molecular weight amides, α fragmentation is predominant.

$$R-\underset{\underset{O}{\|}}{C}-NH_2\rceil^{+\cdot} \xrightarrow{-R^\cdot} \overset{+}{O}\equiv C-NH_2 \longrightarrow O=C=\overset{+}{N}H_2$$

If R contains three or more carbon atoms and if N, in the γ position, loses an H atom, cleavage in the β position by the McLafferty rearrangement gives a prominent peak at $^m/e$ 59 ($CH_2 = \underset{.+\,OH}{C} - NH_2$).

The long chain amides also break in the γ position ($^m/e$ 72) and when followed by a H transposition give an ion at $^m/e$ 73.

There are peaks at M−16 ($-NH_2$) obtained by α cleavage with respect to the carbonyl. Secondary and tertiary amides show a double α cleavage and a C−N cleavage with H rearrangement.

$$R-\underset{\underset{O}{\|}}{C}-NH-R_1\rceil^{+\cdot} \longrightarrow R-\underset{\underset{O+}{\|}}{C}-NH_2 \qquad R-\underset{\underset{O}{\|}}{C}-\overset{+}{N}H_3$$

Nitro-compounds. A very intense peak is derived by elimination of NO_2. The loss of NO occurs frequently and is evidence for the isomerization of the nitro group.

$$C_6H_5-NO_2\rceil^{+\cdot} \longrightarrow C_6H_5-ONO\rceil^{+\cdot} \xrightarrow{-NO^\cdot} C_6H_5O\rceil^{+} \xrightarrow{-CO} C_5H_5^+$$

At the same time there is an intense peak at mass $^m/e$ 30 (NO^+).

8. Halogenated Derivatives.

The halogens, having a great affinity for electrons, do not show a marked tendency to carry the positive charge.

The aliphatic compounds lose the halogen or the hydrohalide. Similarly, the loss of halogen is preferred in the aromatic derivatives.

The compounds are easily recognized in the mass spectrum by the isotopic peaks of the bromo- and chloro-derivatives.

9. Sulphur Derivatives.

They are recognized by their isotopic abundances.

Thiols. The molecular peaks are visible. The $(M-H_2S)$ peak is strong. There is a series of peaks at masses 47 + 14n.

Sulphides. The molecular peaks are visible. The (M−34) peak is absent, and therefore these compounds can be differentiated from their isomers, the thiols.

10. Heterocycles.

The heterocycles such as furan, thiophene, pyrrole, pyridine, and indole behave like the benzene ring, showing an intense molecular peak.

Side-chains preferentially fragment in the β position with respect to the ring.

F. HANDY RULES OF FRAGMENTATION.

The following observations are aids to the identification of parts of unknown structures. They are not absolute rules, and exceptions are frequently found.

1. The relative height of the molecular peak decreases with increased branching, in an homologous series.
2. The relative height of the molecular peak increases if there are double bonds or rings (in particular, aromatic rings).
3. Fragmentation at the most substituted carbon atom is favored.
4. In the case of double bonds, allylic cleavage occurs (β cleavage with respect to the double bond).

$$[CH_2{=}CH{-}CH_2 \vdots CH_2{-}R]^{+\cdot} \xrightarrow{-\dot{C}H_2R} CH_2{=}CH{-}\overset{+}{C}H_2 \longleftrightarrow \overset{+}{C}H_2{-}CH{=}CH_2$$

5. The saturated ring tends to lose the side-chain by α cleavage.

$$[C_6H_{11}{-}CH_2{-}R]^{+\cdot} \longrightarrow C_6H_{11}^{+} + \dot{C}H_2R$$

6. Aromatic compounds with alkyl substituents give rise to β cleavage, with respect to the ring, and the formation of the tropylium ion $C_7H_7^+$.

$$C_6H_5\text{-}CH_2R^{+\cdot} \longrightarrow C_7H_7(H)(R) \xrightarrow{-R^\cdot} C_7H_7^+$$

7. In the presence of a heteroatom, the cleavage occurs preferentially in the β position, and the charge is carried by the fragment containing the heteroatom.

$$[R\text{-}CH_2 \vdots CH_2\text{-}OR]^{+\cdot} \xrightarrow{-RCH_2^\cdot} \overset{+}{C}H_2\text{-}OR \longleftrightarrow CH_2\text{=}\overset{+}{O}R$$

CHAPTER V

A. INTRODUCTION TO MASS FRAGMENTOGRAPHY.

The fundamental equation of a scanning magnetic mass spectrometer, as given in Chapter 1, is

$$\frac{m}{e} = \frac{H^2R^2}{2V} \qquad (1)$$

This equation expresses the relationship between the mass-to-charge ratio ($^m/e$) of an ion, the intensity of the magnetic field (H), the radius of the trajectories of the ions (R), and the applied accelerating voltage (V).

When e=1 and H is constant, equation (1) becomes

$$m = \frac{k}{V} \qquad \text{where } k = \tfrac{1}{2}H^2R^2 \qquad (2)$$

The mass of the ion focused on the collector of the mass spectrometer is therefore inversely proportional to the applied accelerating voltage, and this relationship suggests that the different ions can be focused by varying the accelerating potential, at constant magnetic field. By selecting the appropriate values of the voltage, the desired ions can be focused. By suitably varying the accelerating voltage between certain fixed mass numbers, the ions corresponding to these preselected values can be focused, successively. The time interval between each focalization is so short that the process can be considered as simultaneous.

This, in essence, is mass fragmentography (MF), also known as the single ion detection (SID) or multiple ion detection (MID) technique. Mass fragmentography is therefore the simultaneous detection of one or more fragment ions, as opposed to the scanning of the whole spectrum that occurs in conventional mass spectrometry.

Gas chromatography (GC) is an important technique that has been developed to achieve higher degrees of separation of compounds in a mixture. There are other chromatographic methods (such as thin layer, liquid, and paper chromatography, and electrophoresis), but the gas chromatograph is the easiest and the most applicable system coupled to the mass spectrometer, and for this reason it has been widely used.

Two or three peaks in the gas chromatogram may be only partially resolved, if not completely superimposed. In these conditions it is impossible to study the corresponding mass spectra because the compounds are not pure. The problem can be overcome by repeated scanning during the elution and by reconstructing the elution graph of the individual compounds. This procedure requires time and patience or the use of a sophisticated "on-line" computer.

For this reason, a relatively simple artificial technique, an accelerating voltage alternator (AVA), is applied to the mass spectrometer. This can continuously focus one, two, or three characteristic ions onto the collector by automatically varying the accelerating voltage. It can perform this task with such rapidity that the different ions can be considered to be displayed almost simultaneously.

In practice the magnetic field is adjusted manually until the lower mass is focused, then, while a constant field is maintained, the value which focuses the second fragment is selected on the

AVA. A similar procedure is performed for the third fragment. This method is based on the relationship

$$m \propto \frac{1}{V} \tag{3}$$

For example, figure 27 shows the GC trace of two structurally similar compounds, A and B. These molecules have very similar retention times on the GC column and therefore can only be partially resolved.

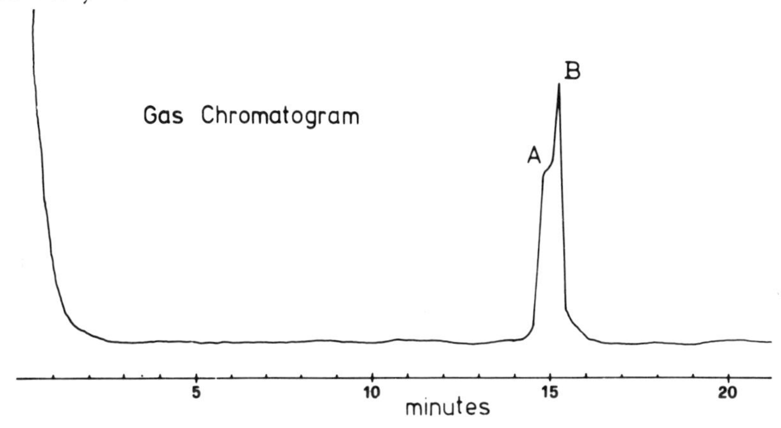

Figure 27 Gas chromatogram of two unresolved compounds, A and B.

On electron bombardment, A gives a fragment ion at mass 270, whereas B produces a similar fragment that has a mass 272. Fixing the AVA at masses 270 and 272 the two substances can be resolved (figure 28).

B. PHARMACO-BIOLOGICAL APPLICATIONS OF MASS FRAGMENTOGRAPHY.

With few exceptions mass fragmentography, as a single or multiple ion detection method applied to the gas chromatograph, has been used principally in the field of quantitative and qualitative analysis of pharmacologically active substances.

Some of these compounds, particularly the hallucinogens, are

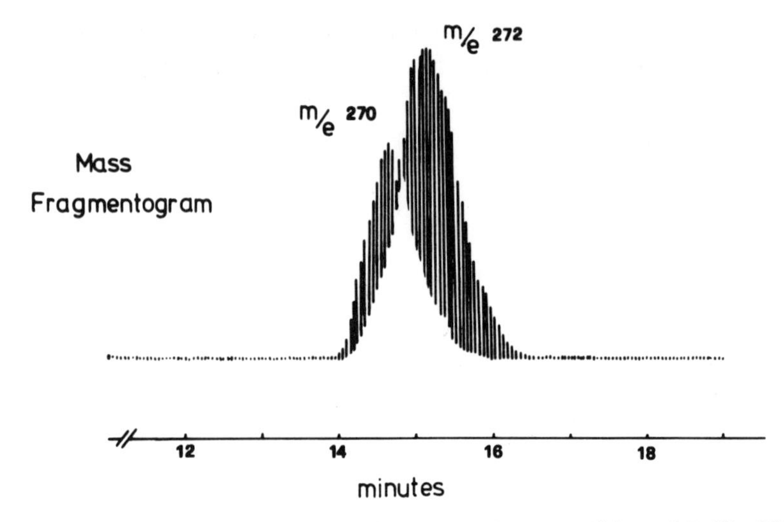

Figure 28 Mass fragmentogram of compounds A ($^m/e$ 270) and B ($^m/e$ 272).

active in the order of microgram quantities, and therefore their identification from tissues or biological liquids requires the use of a very sensitive method, such as mass fragmentography.

The numerous applications of this technique can be conveniently sub-divided into two categories—detection of a single ion (SID), and detection of multiple ions (MID).

C. Single Ion Detection.

In single ion detection, the mass spectrometer is used as a selective gas chromatographic detector. It is not necessary to use the AVA in these cases. Only one ion is focused (generally an intense one), so that only when the compound begins to be eluted from the GC column, even in the presence of a complex mixture, there is a trace on the oscilloscope.

1a. Selective Gas Chromatographic Detector.

This technique can increase the sensitivity of detection by a factor of between 1,000 and 10,000 with respect to normal gas chromatographic detectors, such as electron capture, flame ionization, or thermal conductivity. Therefore a substance that would not appear on a conventional GC trace can be recorded.

An example of this application is the determination of STP (2,5-dimethoxy-4-methyl-amphetamine; *S*erenity, *T*ranquillity, and *P*eace), a potent hallucinogen. Figure 29 shows its mass spectrum with a prominent peak at m/e 166.

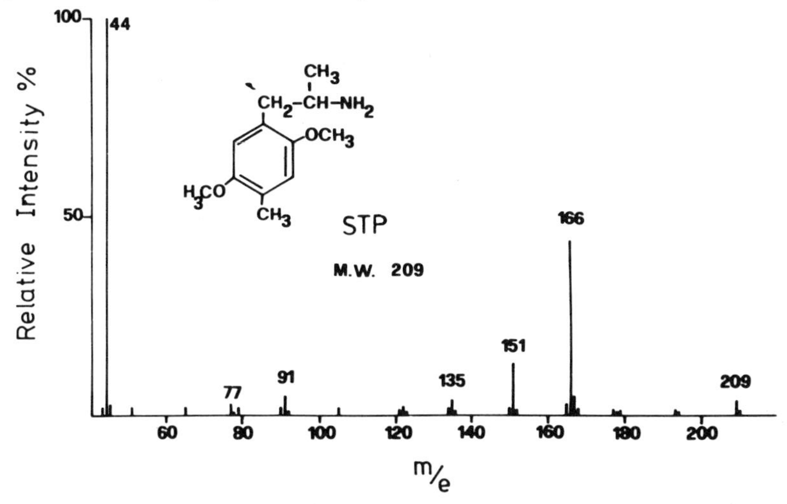

Figure 29 Mass spectrum of 2,5-dimethoxy-4-methyl-amphetamine (STP).

In figure 30 the gas chromatogram obtained by recording the total ion current is shown. The mass spectrometer was used as the detector. The retention time of STP, in isothermal conditions at 140°C. on a 3% OV 17 glass column, is 5 minutes, but the peak was not observed when the amount of sample was in the order of 1 ng.

Figure 31 shows the mass fragmentogram of STP under the same conditions. The mass spectrometer is employed to detect a single ion and is focused on mass m/e 166. Although mass m/e 44 is more abundant, it is not used in this case because it is not characteristic of STP alone.

The sensitivity of this method is such that STP can be detected in quantities as low as 100 pg. DDT, in trace amounts as low as 10 pg., has been similarly determined.

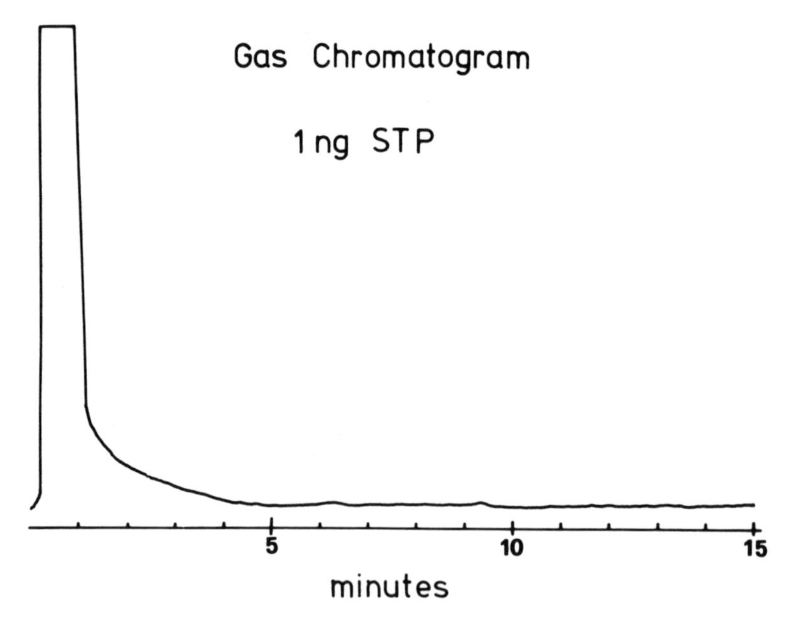

Figure 30 Gas chromatogram of 1 ng of STP.

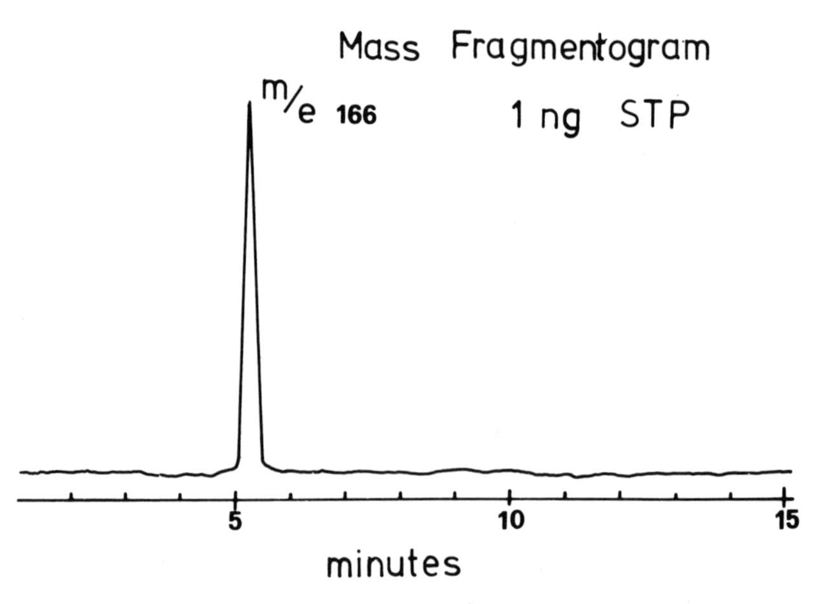

Figure 31 Mass fragmentogram of 1 ng of STP by focusing on $^{m}/e$ 166.

1b. Single Ion Detection and High Resolution Mass Spectrometry

Combining the single ion detection method and a high resolution instrument with a resolving power in the order of 100,000 it is possible to distinguish between ions that have the same nominal $^m/e$ ratio but have a different elemental composition.

The method has been applied to the determination of p-tyramine, a precursor of catecholamines, isolated from brain extracts. The sample was introduced directly into the ion source, rather than by way of the gas chromatograph, and a doublet was observed at the nominal mass 108. The high resolution mass spectrometer was focused on $^m/e$ 108.0575, an ionic fragment corresponding to p-tyramine, permitting the compound under examination to be distinguished from interfering lipid hydrocarbons that have a peak at $^m/e$ 108.0939.

1c. Multi-compound Monitoring (Detection of Compounds of the Same Series by SID).

The SID can be conveniently employed for compounds of the same series having different retention times on the gas chromatogram but producing the same fragment ion by electron bombardment.

This technique has been used with good results in the selection of the aryl GC peaks out of a mixture containing alkyl and aryl compounds. This, in fact, was the first example of the use of mass fragmentography as a single ion detector.

When the instrument is focused on ion $^m/e$ 91, the tropylium ion formed from aryl compounds by β-cleavage, only this series can be detected by the recording oscillograph. The overlapping of the gas chromatographic and the oscillographic peaks (fragmentogram) indicates the aryl compounds (figure 32).

The same method can be employed for displaying the paraffins, alcohols, esters, and acids.

The estrogens have a common, fundamental structure and by electron impact produce a peak at $^m/e$ 216. When the mass

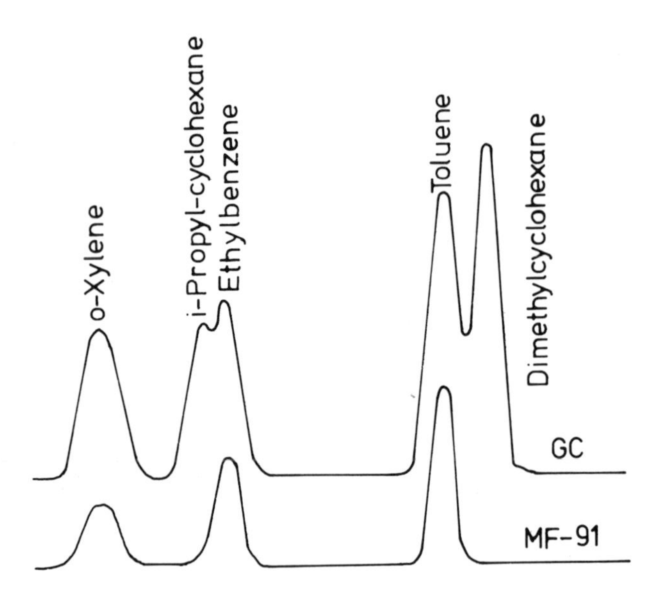

Figure 32 Gas chromatographic (GC) and mass fragmentographic (MF) recording of a mixture of compounds, focusing the instrument on mass $^m/e$ 91.

spectrometer is focused on this value, a trace on the oscillograph is observed that is coincident with the trace of 17β-estrenol and its 17α-analogues from the gas chromatogram.

2. Multiple Ion Detection.

When the mass spectrometer is employed as a multiple ion detector it focuses on two or three ionic fragments within a 10–30% range of the magnetic scan of the instrument, or on up to eight fragments over the whole field using a quadrupole mass spectrometer.

For example, the mass spectrum of desmethylchlorpromazine trifluoracetate (DMCP-TFA) shows three prominent peaks, at $^m/e$ 232, 234, and 246 (figure 33).

Figure 34a shows the trace recorded on the oscillograph of a fraction eluted from the gas chromatograph and containing the desired compound. The trace was obtained from an instrument having an AVA and focusing on the three peaks at $^m/e$ 232, 234,

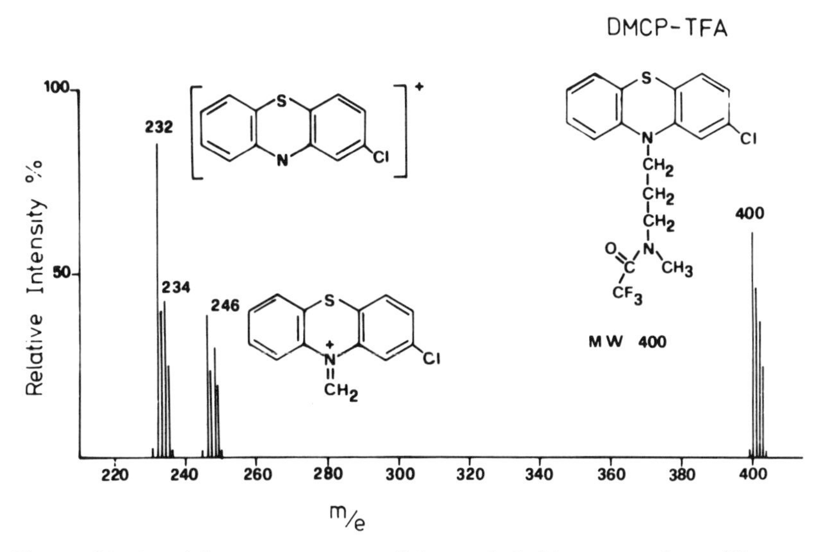

Figure 33 Partial mass spectrum of desmethyl chlorpromazine trifluoracetate (DMCP-TFA).

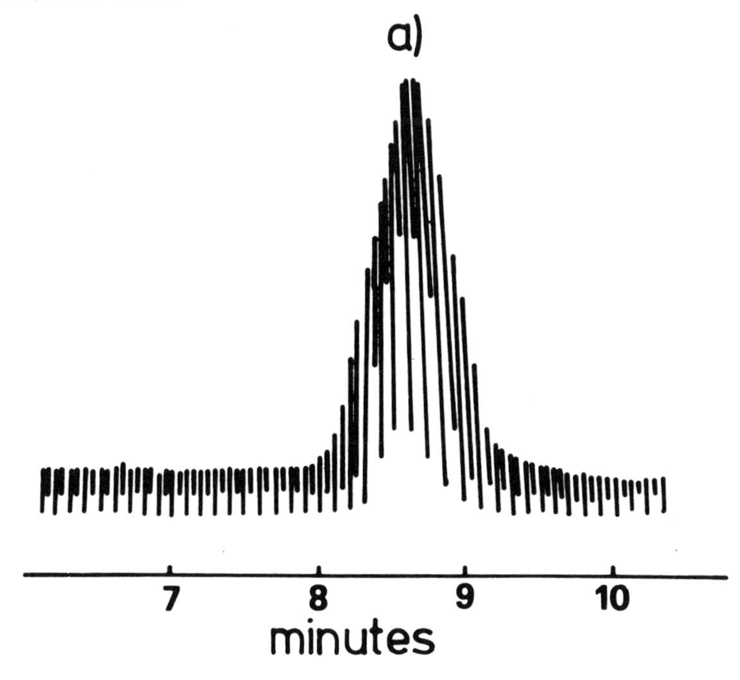

Figure 34 (a) Mass fragmentographic trace of DMCP-TFA.

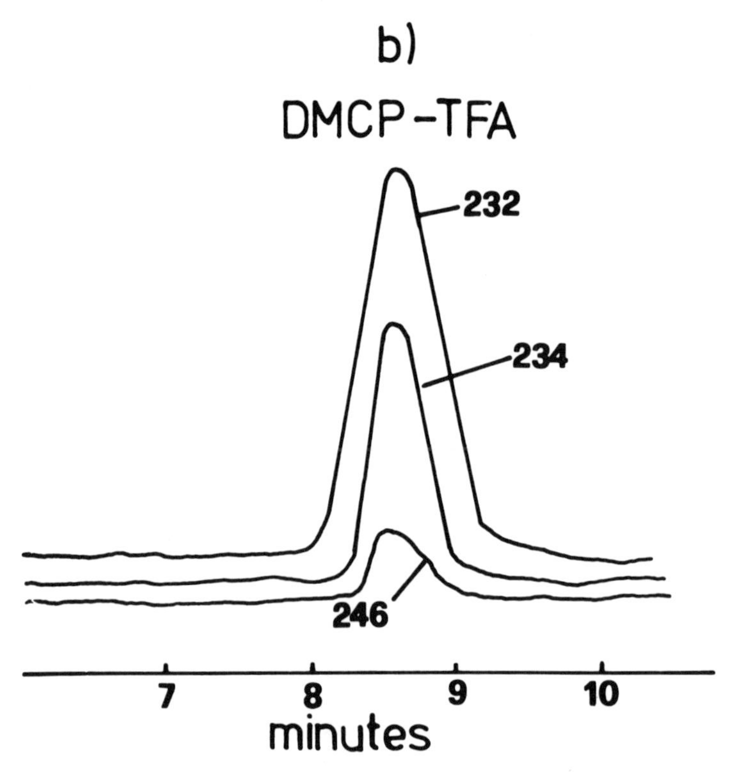

Figure 34 (b) Graphic representation of the fragmentogram.

and 246. The graphic form (figure 34b) is a clearer representation of the fragmentogram.

2a. Construction of Spectra from Successive Fragmentograms.

When a compound can only be obtained in minute quantities, too low to allow a complete recording of the mass spectrum in the conventional manner, successive recording of the partial mass fragmentograms, obtained by focusing on three spectral lines and refocusing every time on different groups of peaks, one can construct a complete mass spectrum.

This technique has been employed with success in the construction of complete mass spectra of some metabolites of chlorpromazine.

2b. Research into Precursors or Metabolites.

Some precursors and metabolites of natural or synthetic substances can be hypothesized *a priori* on the basis of biochemical knowledge. If a chemical structure is postulated for such compounds, the molecule can be synthesized and its mass spectrum recorded *or* the expected ionic fragments foreseen by relying exclusively on characteristics in a fragmentation scheme.

In both cases the mass numbers, which focus the instrument to detect the desired compound, have to be sought.

This method has been successively applied in the identification of 10-hydroxy-nortriptyline, desmethyl-nortriptyline, and desmethyl-10-hydroxy-nortriptyline, which are metabolites of nortriptyline (a widely used antidepressant drug). It has also been used in the determination of the precursors of the mescaline hallucinogens and tetrahydro-isoquinoline.

2c. Quantitative Evaluation at Sub-nanogram Levels with High Specificity.

This method is analogous to those illustrated above using the single ion detection method, but has the additional advantage of greater specificity. This result is obtained by focusing on more than one characteristic ionic fragment of a molecule and this ensures the identity of the compound displayed. One of the channels on the AVA can also be used to detect the intensity of a fragment ion derived from an internal standard. This is another application of mass fragmentography.

Figure 35 shows how this method was used in the quantitative determination of imipramine, an antidepressant drug. The spectrometer was focused on fragments at $^{m}/e$ 235 and 238, characteristic of imipramine (IMI) and promazine (PRO) respectively. PRO was introduced in a known quantity as internal standard. The quantitative estimation was given by the ratio of the relative peak areas of the compound and the internal standard and by comparison with a calibration curve.

One of the advantages of this method is that the estimation of a substance is simultaneously qualitative (identification of the substance) and quantitative (extremely precise estimation of the

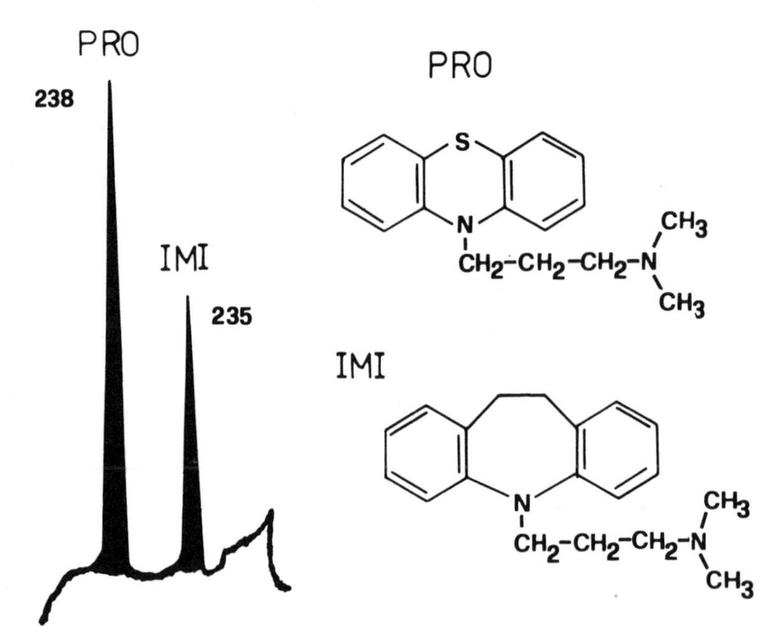

Figure 35 Multiple Ion Detection of imipramine (IMI) and its internal standard promazine (PRO) with the instrument focused on fragments $^m/e$ 235 and 238.

quantity, even if very small). Even in the presence of complex mixtures, as in extracts from biological fluids, the compound can be detected without interference from the other components in the system. Interference can sometimes come from the GC column by the "bleeding" effect.

2d. Multiple Ion Detection and Stable Isotopes.

For quantitative measurement, in metabolism studies, the reference compound should ideally be chemically similar to the compound under examination and detectable by the same method. Therefore, derivatives or compounds labelled with ^{15}N or ^{2}H are suitable.

Labelled analogues. Analogues of the compounds under study, labelled with ^{15}N or ^{2}H, show the same fragmentation scheme as the unlabelled molecules. Nevertheless, after fragmentation the fragments derived from the labelled compounds will be easily distinguishable from the unlabelled fragments because of their higher mass numbers. Analogues, suitably labelled, are therefore used for quantitative measurement and the estimation of the percentage recovery of endogenous compounds.

Derivatives. Many compounds can be derived before being subjected to gas chromatographic analysis. A suitable internal standard can be obtained by performing the derivation with suitably labelled reagents. A ^{2}H labelled reagent such as $^{2}H_3$-O-methyl-hydroxylamine can form the $^{2}H_3$-methoxy derivative of the reference compound. The equivalent unlabelled reagent is employed in the treatment of the biological compound. By focusing the spectrometer on the fragments of mass M of the derived compounds and on mass M+3 for the internal standard, accurate quantitative estimations of the endogenous levels of biological compounds can be obtained.

This method has been elegantly applied in the estimation of prostaglandin E_1.

Numerous deuterated compounds are available and can be used to derive alcohols, phenols, amines, etc.

Naturally Occurring Stable Isotopes. 25% of naturally occurring chlorine has an atomic weight two mass units higher than that of the most abundant isotope, ^{35}Cl. Therefore, all molecules and fragments containing chlorine give rise to two ions in the ratio 3:1, separated by two mass units. By focusing the instrument simultaneously on the two fragments M and M+2, a high specificity can be obtained, as demonstrated in the case of chlorpromazine.

Figure 36 shows the GC trace and the mass fragmentogram of chlorpromazine (obtained by focusing on ions 318 and 320). The difference in mass is due to the chlorine isotopes. Furthermore, the molecular ions are in the ratio 3:1. The sensitivity of the method is such that a compound in the order of 1 pg can be detected.

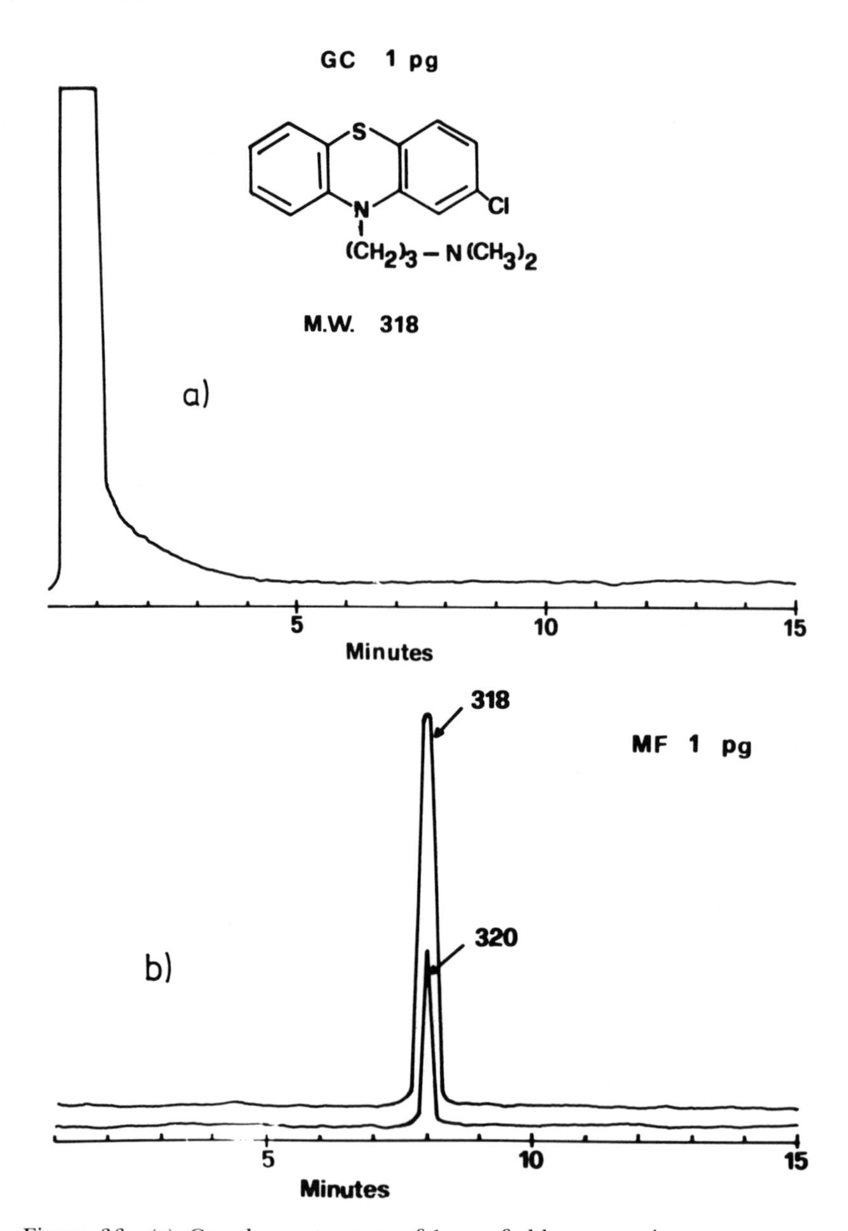

Figure 36 (a) Gas chromatogram of 1 pg of chlorpromazine.
(b) Mass fragmentogram of 1 pg of chlorpromazine with the instrument focused on masses $^{m}/e$ 318 and 320.

Chlorpromazine is a drug often used in the treatment of mental disease and has been employed since 1952. The identification of its metabolites in the plasma of treated subjects, by means of gas chromatography–mass spectrometry, presented many difficulties because of the low concentration of such compounds. In fact, chlorpromazine is rapidly metabolized by an organism, and only low levels remain in the plasma, even if the dose administered was relatively high.

The GC-MS spectra of chlorpromazine and some of its chemically synthesized, hypothetical metabolites, such as monodesmethyl- and didesmethyl-chlorpromazine, have been studied. During electron bombardment these molecules fragment by common routes. In the mass spectra of these compounds a fragment at $^{m}/e$ 232 is always present, due to the chloro-phenothiazinyl ion. Another at $^{m}/e$ 246 is due to the ion with an N-methylene group in the phenothiazine (figure 37). Because of the presence of both chlorine and sulphur, the isotopic ion at mass M+2 was 40% as intense as the ion of mass M.

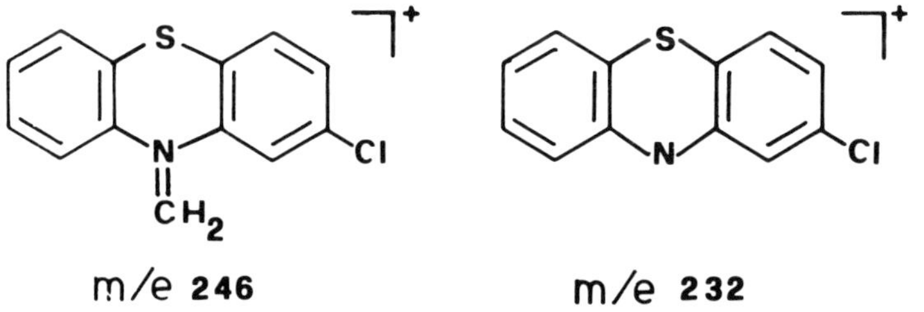

m/e **246**

m/e **232**

Figure 37 Fragment ions of chlorpromazine, after electron bombardment.

The various fragments present in the mass spectrum of these substances have different relative abundances, and therefore the various compounds can be distinguished from one another.

If one knows the fragmentation pattern of these molecules, the presence of chlorpromazine metabolites can be demonstrated, by focusing the spectrometer on the three masses $^{m}/e$ 232, 234, and 246.

This method was used to analyze the metabolites contained in the biological fluids of certain patients who had received 75 mg/day of chlorpromazine.

When the mass spectrometer is used as a multiple ion detector, the presence of the drug and its metabolites can be determined on the basis of their retention times on the gas chromatogram and the corresponding mass fragmentogram.

Because these substances were present in quantities in the order of 10 pg (10×10^{-12}g), no other analytical technique would have been able to detect them.

Thus, in the field of pharmaco-biological research, this technique is invaluable for determining the identity and the quantity of chemical compounds in biological material. The method is more accurate and more sensitive than any other method available and if coupled to a computer system it could make the quantitative determination of drugs and their metabolites even simpler.

C. GAS CHROMATOGRAPH-MASS SPECTROMETER-COMPUTER.

In order to contend with the numerous spectra obtained by GC-MS analysis of a mixture of compounds, instruments have been devised to computerize the mass spectral data.

During the analysis of a natural, complex molecule it is possible to record hundreds of spectra, and these spectra must be correlated with the gas chromatographic peaks by an efficient number system. Moreover, every single spectrum must be examined, the characteristic fragment peaks must be identified, and their intensities measured. Any peak due to the background or to column "bleeding" must be subtracted and the spectrum normalized. The normalization is performed by displaying, in graphic form, each fragment as a percentage of the most intense fragment found in the spectrum.

Then follows the interpretation and the identification of the structure of the compounds present.

It is obvious that the accumulation of data from a single scan is very laborious, and several attempts have been made to use a computer system—for example, manual analysis using analog tapes and an "off-line" computer, manual analysis using digital output and an "off-line" computer, computer data logging, and acquisition of data by an "on-line" computer that can control the mass spectrometer.

The computer (on-line) is commanded by a simple dialogue through a teletype. The operator punches in the necessary requirements: the title (for the file), the mass range (first and last mass), the integration time (in milliseconds), and the number of seconds per scan.

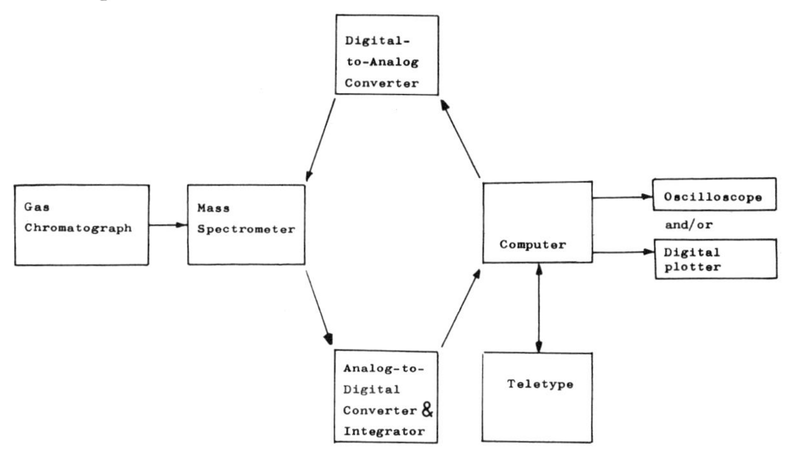

Figure 38 Scheme of a gas chromatograph-mass spectrometer-computer system.

At the interface between the computer and the mass spectrometer there is a digital-to-analog converter (figure 38), which takes the digital information punched in by the operator and converts it into an analog voltage to the magnetic analyzer or to the quadrupole rods.

The sample that has passed through the gas chromatograph enters the ion source and becomes ionized. The total ion current is monitored as the ions pass through the slit and enter the magnetic analyzer. The total ion current is displayed on the oscilloscope and resembles the gas chromatogram.

Mass spectra are obtained regularly depending on the

number of seconds per scan, and the integrated output is converted into digital form by an analog-to-digital converter. The data is transferred to the computer, which normalizes the spectra and can display them on the oscilloscope or store them on magnetic tapes or discs.

A timing generator automatically regulates the integration time as a function of the signal intensity. This method of optimization of the signal gives an almost constant signal-to-noise ratio over the whole spectrum.

There are numerous accessory programs (software) that can be used to calibrate the spectrometer automatically using a standard, control the instrument during the scan, acquire the data, display the data in printed or graphic form, and conduct a library search.

APPENDIX

A. COMMON SYMBOLS AND ABBREVIATIONS.

Abbreviations.

$A(R_1)$	Appearance potential of ion R_1^+
$I(R_1R_2)$	Ionization potential of molecule R_1R_2
$^m/e$	Mass-to-charge ratio
Rel. Int.	Relative intensity
a.m.u.	Atomic mass unit ($^1/_{12}$ mass of the most abundant isotope of carbon)
eV	Electron volt
M.W.	Molecular weight

Symbols.

$M^{+\cdot}$	Molecular ion
$(X)^+$	An ion with an even number of electrons. For example, $(M-CH_3)^+$
$(X)^{+\cdot}$	An ion with an odd number of electrons. For example, $(M-CO)^{+\cdot}$
m*	metastable peak
m* 43→28	metastable transition from ion 43 to ion 28

(calc. 18.2, obs. 18.3) metastable peak observed at $^{m}/e$ 18.3, calculated value 18.2

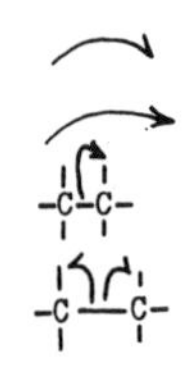

Transfer of *one* electron

Transfer of *two* electrons

Heterolysis

Homolysis. For clarity only one arrow is shown. An example: $R-C(=O^{+\cdot})-R \rightarrow R^{\cdot} + {}^{+}O \equiv C-R$

$-CH_2 \vdots CH_2-$ 58:69 — Fragmentation of a single bond giving a fragment of mass 58 and another of mass 69

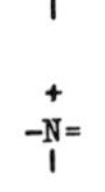

Radical ion formed by extraction of an electron from nitrogen

An ion with an even number of electrons and the charge located on nitrogen

$R]^{+}$, $R]^{+\cdot}$ — Nonlocalized charges

$15\% \Sigma_{40}$ — 15% of the total ion current calculated from ion $^{m}/e$ 40 to the molecular peak

x 10 — Indicates a peak enlarged by a factor of 10

B. DEFINITIONS.

Molecular ion	The ion formed by the loss of one electron from the original molecule.
Molecular cation	Same as above. Used when there are positive and negative ions in the same spectrum.
Molecular anion	An ion formed by the capture of an electron by the original molecule.
Rearranged molecular ion	An ion having the same mass but not the same structure as the molecular ion.
Fragment ion	An ion formed by the cleavage of one or more bonds of the molecular ion. It can be an ion with either an odd or an even number of electrons.

Rearrangement ion	An ion formed by a rearrangement process.
Parent peak	The origin of a fragment ion.
Base peak	The most intense peak in the spectrum. It is given an arbitrary intensity of 100.

ESSENTIAL BIBLIOGRAPHY

A. BOOKS.

Beynon, J. H. (1960) *Mass Spectrometry and Its Applications to Organic Chemistry,* Elsevier.

Beynon, J. H., Saunders, R. A., and Williams, A. E. (1968) *The Mass Spectra of Organic Molecules,* Elsevier.

Biemann, K. (1962) *Mass Spectrometry: Organic Chemical Applications,* New York, McGraw-Hill.

Budzikiewicz, H., Djerassi, C., and Williams, D. H. (1964) *Structure Elucidation of Natural Products by Mass Spectrometry.* v. 1, *Alkaloids;* v. 2, *Steroids, Terpenoids, Sugars, and Miscellaneous Classes.* Holden-Day.

Budzikiewicz, H., Djerassi, C., and Williams, D. H. (1967) *Mass Spectrometry of Organic Compounds,* Holden-Day.

Burlingame, A. L., ed. (1970) *Topics in Organic Mass Spectrometry,* Wiley Interscience.

Frigerio, A., and Castagnoli, N., eds. (1974) *Mass Spectrometry in Biochemistry and Medicine,* Raven.

Hill, H. C. (1966) *Introduction to Mass Spectrometry,* Heyden and Son.

Knewstubb, P. F. (1969) *Mass Spectrometry and Ion-Molecule Reactions,* Cambridge University Press.

McLafferty, F. W., ed. (1963) *Mass Spectrometry of Organic Ions,* Academic Press.

McLafferty, F. W. (1967) *Interpretation of Mass Spectra,* Benjamin.

Melton, C. E. (1970) *Principles of Mass Spectrometry and Negative Ions,* M. Dekker.

Milne, G. W. A., ed. (1971) *Mass Spectrometry Techniques and Applications,* Wiley Interscience.

Roboz, J. (1969) *Introduction to Mass Spectrometry: Instrumentation and Techniques,* J. Wiley and Sons.

Spiteller, G. (1966) *Massenspektrometrische Strukturanalyse Organischer Verbindungen,* Verlag Chemie.

Waller, G. R., ed. (1972) *Biochemical Applications of Mass Spectrometry,* Wiley Interscience.

Williams, D. H., and Fleming, J. (1966) *Spectroscopic Methods in Organic Chemistry,* McGraw-Hill.

B. COLLECTIONS OF SPECTRA.

American Society for Testing and Materials (1969) *Index of Mass Spectral Data.*

Cornu, A., and Massot, R. (1966) *Compilation of Mass Spectral Data,* Heyden and Son.

Mass Spectrometry Data Centre (1970) *Eight Peak Index of Mass Spectra,* Aldermaston.

Mass Spectrometry Data Centre, *Mass Spectrometry Data Centre Series of Mass Spectral Data,* Aldermaston.

Stenhagen, Abrahamsson, McLafferty, eds. (1969) *Atlas of Mass Spectral Data,* Interscience Publishers.

SUBJECT INDEX